THE NEW WORLD CHAMPION
PAPER AIRPLANE BOOK

THE NEW WORLD CHAMPION PAPER AIRPLANE BOOK

FEATURING THE **WORLD RECORD**– BREAKING DESIGN

—

WITH TEAR-OUT PLANES TO FOLD AND FLY

JOHN M. COLLINS

TEN SPEED PRESS
Berkeley

To my parents, Ted and Marie, for teaching me how to enjoy little, everyday things. They encouraged me to dream, and helped me learn how to learn.

Published in the United States by
Ten Speed Press, an imprint of the
Crown Publishing Group, a division
of Random House, Inc., New York.
www.crownpublishing.com
www.tenspeed.com

Ten Speed Press and the Ten Speed Press colophon are
registered trademarks of Random House, Inc.

Library of Congress Cataloging-in-Publication Data
is on file with the publisher.

Trade Paperback ISBN: 978-1-60774-388-0
eBook ISBN: 978-1-60774-389-7

Printed in the United States of America

Design by Katy Brown

10

First Edition

CONTENTS

INTRODUCTION

On February 26, 2012, my paper airplane design, Suzanne, flying 226 feet and 10 inches, broke the old paper airplane world record for distance by 19 feet and 6 inches. It's fair to say that Joe Ayoob, my thrower, and I shattered it. The old mark had stood for nine years. Joe and I had made better distances in the six prior practices, but then, that's what makes record breaking so interesting. A record during practice isn't a record. It's not official until declared so under the specific guidelines during a sanctioned attempt.

Records are made to be broken. It's the whole reason for having them. The goal, the chase, the falling short, and the trying again are all required components of a record attempt. This record was no different. I tried many models over a three-year period when I was "officially" working on breaking the record. Of course, the truth is, every great distance plane I ever made was step toward this goal. My hope is that your world-record journey starts, or continues, here. Helping someone else break this record is another of my goals. Perhaps it will be you.

Believing you can do it is the first step. Making something as sophisticated as a flying machine from the most modest of resources is the beginning of an adventure. Where does it lead? Using less to make more is where we have to go as a planet. Eventually we'll find a way to use less energy to light a room, less fuel to move from home to job, and make less pollution from powering the world. Viewed from this perspective, the answers seem obvious: conservation—yes, creating new products and technologies—yes, doing less—no, doing more with less—yes. And we're back to paper airplanes.

I'm frequently asked what advice I have for budding paper pilots. Oddly, I'm usually at a loss. Folding comes so naturally for me, I can't imagine *not* making paper airplanes. It's taken me years to figure out that not everyone is like this. It's true that some people might actually need encouragement. After giving this due consideration, I've come up with the following suggestion.

If you've never been to a Maker Faire, find one and go. It's a celebration of making things—all kinds of things—from tiny robots to clothes to giant sculptures of steel and stone, and yes, sometimes paper airplanes. Making things is part of who we are as Americans and who we are as humans. From the biological imperative to make more people to the need to know who made that cake, we're hardwired to like the whole idea.

As a kid, I liked making toys. My brothers and I would make spinning tops from wooden spools, rubber-band guns using spring-loaded clothes pins as the triggers, parachutes from napkins and string, rubber-band powered boats, kites, balsa-wood planes, and yes, paper airplanes.

Sadly, making things has been largely lost as part of our culture. Toys now need to "do something" so that even playing with the thing is optional. In my view, this is a huge mistake. We are robbing ourselves of a very important experience: experimenting, exploring, creating—in short, *making things*.

Paper airplanes embody the scientific method. Every throw is an experiment. It's a hobby that begs the paper pilot to understand ever more in order to excel. Hypothesis, experiment design, trial, and results—it's all built into every plane and every throw. To play with a paper airplane is to dabble in science, whether you know it or not.

We have a number of global issues confronting us. Global energy shortages, food shortages, water shortages, and something people are calling global warming are all worrisome. These problems will have answers that only science can provide. We have no spare brains on the planet. We need everyone thinking about these challenges in a rigorous way.

Imagine this: a world of people playing with science, who get up every morning, focus on what's good, and imagine how to make more of that. You can call me a dreamer. I don't mind. You don't have

to believe a word of what I say. Just make a paper airplane and experience how exhilarating that feels. We're born makers. When you make something, anything from a pie to a pencil drawing, it's like waking a dormant part of you. The world shifts slightly. You can feel it, and it feels good.

Suzanne, the world-record paper airplane, boasts a series of firsts: the first glider to hold the distance record, the first paper airplane to use changing airspeed to enhance performance, the first plane to use a thrower/designer team, and the first plane to break the record after the run-up-to-throw distance was shortened from 30 to 10 feet. It is a truly amazing aircraft. I believe Suzanne changes the way distance records will be broken in the future. The days of brute-force darts are gone, replaced by the age of true gliders.

A little free advice: take nothing for granted. Suzanne is a great aircraft. I didn't find the design hiding in someone else's work. I created it by working hard, listening closely, and observing keenly. You probably possess these skills too. This plane is only one of many possible solutions to the challenge. You may discover others.

Fold an extra plane for me, and perhaps I'll meet you in the winner's circle.

WHY STUFF FLIES

Here's the short version: we're not sure why stuff flies. And in "we," I'm including people who've spent more time studying the matter than I have. I'm an intellectual interloper. I don't have an aeronautics degree, just a high degree of curiosity. That "we're not sure" answer, however, could stand some expansion—at least in relation to paper airplanes, which is what we'll look at here.

Paper airplanes are **GLIDERS** because they, well, glide. In short, they have no motor. Let's take a look at some of the dynamics or **FORCES** that make paper airplanes glide.

BASIC FORCES

Let's start with what everyone can agree on: The most basic forces involved in paper airplane flight are lift, weight, drag, and thrust. **LIFT** is the upward force generated as a plane moves through the air. **WEIGHT** is the force caused by the gravitational pull of the earth, while **DRAG** is the resistance created by a shape or material that impedes forward motion. **THRUST** is the force supplied by a motor on powered planes. With a glider, thrust is more complex, since the only source of thrust is your initial throw. The energy from that throw is converted into momentum, which will stretch over the whole flight. It's a little like that first high drop from a roller coaster: it has to supply enough momentum to keep the coaster moving throughout the whole trip. Good paper airplanes are designed to withstand a short, fast period of thrust (your throw). Once that thrust is used up, the plane needs to balance the remaining forces of drag, lift, and weight to stay in the air.

Figure 2 on the next page illustrates these four basic forces. Understanding the four basic forces is a useful way to think about paper airplane flight. By defining aircraft design as the best way to balance these forces, thinking about and solving flight problems becomes easier.

Now that we understand the four forces acting upon an aircraft, let's take a look at the basic anatomy of a plane. Figure 1 illustrates the various parts of a powered plane.

If we begin to move elements around, we can easily predict the outcome. For instance, if we move the main wing very far to the rear but leave the motor way out front (as in Figure 3), the center of gravity is now very far in front of the center of lift. A plane configured like this would crash.

However, if we move the motor to the rear *along with* the wing (as in Figure 4), and add a horizontal stabilizer to the front, balance is restored.

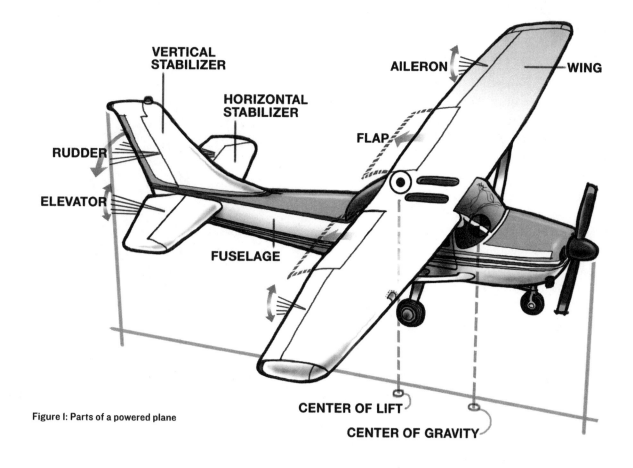

Figure I: Parts of a powered plane

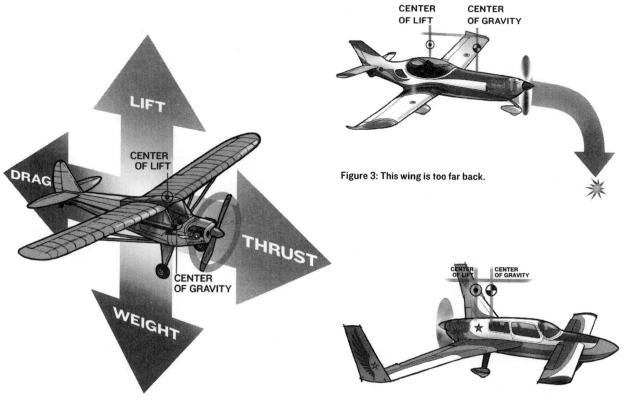

Figure 2: The four forces

Figure 3: This wing is too far back.

Figure 4: The wing and motor are *both* moved back, restoring balance.

WING SHAPE

First, thinner is a winner for paper airplanes. Very flat wings have less drag than thicker or curved wings. This is not true for full-sized airplanes, but more on that later. For paper airplanes, there tend to be two basic wing shapes: rectangular and triangular. The former world-record holder for duration (that is to say, the longest time aloft) had a very rectangular shape. The current world-record holder for duration, on the other hand, is more triangular shaped at the nose, with the tail very rectangular.

One thing to consider in designing a paper airplane is its **ASPECT RATIO**, or, more simply, the ratio of the length of the wing to its breadth. The absolute best shape for a glider is a very long, narrow set of wings—like those of a seagull or albatross. This wing configuration has a **HIGH ASPECT RATIO**—in other words, the distance from wing tip to wing tip is much greater than the distance from the front of the wing to the back. Creating a paper airplane with a wing like a seagull, however, is a tall order for a designer. Paper wings need to withstand a mighty throw and still perform well. High aspect ratio wings are tough to make structurally sound for paper airplanes.

Let's return to our discussion of the basic wing shapes: triangular and rectangular. The advantage to a triangular-shaped paper aircraft is that layers of paper can be moved toward the middle of the plane, which will help with overall stiffening (resulting in a stronger, sturdier plane that will hold up when you throw it). The advantage

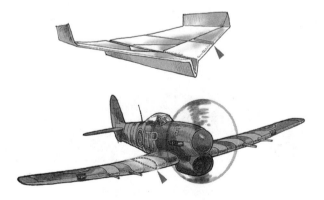

Figure 5: Paper airplane wings should be very thin. Full-sized wings can use thickness to great advantage.

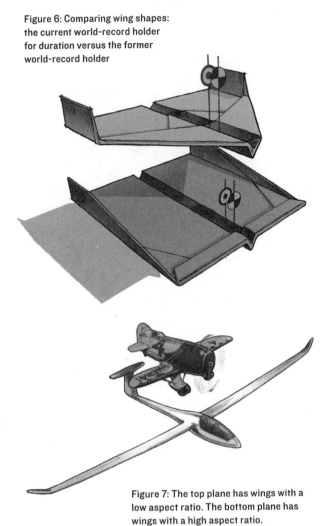

Figure 6: Comparing wing shapes: the current world-record holder for duration versus the former world-record holder

Figure 7: The top plane has wings with a low aspect ratio. The bottom plane has wings with a high aspect ratio.

of a rectangular shape is that it's a more efficient glider, meaning it will fly farther forward for each foot it loses in height.

The shape of a paper airplane determines where its **CENTER OF LIFT** will be. The **CENTER OF GRAVITY** needs to be a bit in front of the center of lift to keep the plane moving forward after your throw. (Refer back to Figure 2 on page 9 for an illustration of center of lift and center of gravity.)

A rectangular-shaped plane needs almost half of the paper weight at or near the nose in order to keep the center of gravity in front of the center of lift. A triangular-shaped plane can have the weight moved further back because there's less wing area (or **LIFTING SURFACE**) at the nose.

This dance between shape and center of gravity makes paper airplane inventing endlessly fun. The tradeoffs work like this: a really efficient glider won't need to be thrown as high to travel far, so a designer could give up some structural support in favor of wider wings (which will allow the glider to stay in the air longer).

A great distance plane needs to be thrown very hard, so more layering will be critical in the leading edges of the wings and the fuselage. Flying in a straight line matters for distance planes, so a taller tail may be the way to go. How much layering? How broad can I make the wings? How tall should the tail be? How hard will I throw? These are all strategic questions for your specific goals. My world-record plane is simply one set of compromises for a goal. Other solutions are certainly possible, perhaps even preferable.

DIHEDRAL ANGLE

Another basic component of flight stability is called **DIHEDRAL ANGLE**, which in airplane design refers to the angle at which the wings are attached to the body of the aircraft. Take a look at Figure 8. If the wings are angled upward from horizontal, they are said to have **POSITIVE DIHEDRAL ANGLE**. If the wings are angled downward from horizontal they are said to have **NEGATIVE DIHEDRAL** or **ANHEDRAL ANGLE**.

Positive dihedral helps put the center of lift above the center of gravity, creating a self-correcting mechanism. If the plane rocks to one side, for

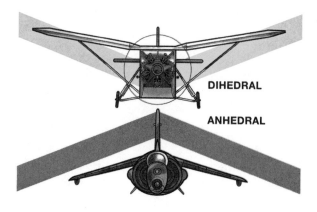

Figure 8: Positive and negative dihedral angles

example, the center of gravity will tend to rock back to vertical. The price for a self-correcting aircraft with positive dihedral angles is a little more drag and a little less lift. Flying is series of negotiations with forces. For passenger planes in particular, the added stability is worth a little less efficiency.

SCALE EFFECTS

Full-sized passenger aircraft and paper airplanes manipulate airflow in different ways. I was able to exploit how the airflow changed on my plane at varying speeds to break the world record. This strategy wouldn't work on a huge 747-sized plane, though.

Figure 9: Scale effects: Think of air molecules like race cars roaring around a track. If they try to take a corner too fast, they can lose traction and won't stay on course.

The dramatically shifting airflow on the paper airplane wing is a product of **SCALE EFFECTS**. Scale effects work like this: Air molecules don't change size or properties. They stay the same, regardless of the wing's size. So when you scale down a 747's gently curving wing to paper airplane size, the air molecules are now making a very sharp turn over a very short distance—just like the race cars in Figure 9 on page 11.

Here's the important concept: there's a limit to how sharp a turn air can make around a fast-moving object—whether that's a 747 or a paper airplane. There's a whole set of numbers, known as Reynolds numbers, that precisely quantify this idea; these numbers measure the ratio of the inertial forces to the viscous forces in a liquid or air. Reynolds numbers allow aerodynamicists to make half-size or quarter-size wings to test in smaller facilities so, for example, they don't have to have a humungous building to test a full-size 747.

What does this mean for paper airplanes? It's better to have a flat, rather than curved wing, so the airflow will be more consistent (and not have to make a sharp turn around a tiny curved wing). During changes in speed, the small wings of a paper airplane can see dramatic shifts in airflow from front to back. The same shift would be much smaller proportionally on a full-sized airplane wing, since the curving is much less abrupt for air molecules on a big wing.

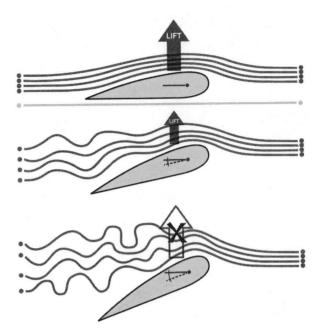

Figure 10: A stall occurs when the angle of attack is too high.

GLIDE RATIO

Glide ratio is a number that's used to describe how far forward an aircraft can travel before losing a standard unit of altitude. Usually, this is expressed as a ratio like 4:1 or 50:1. In those examples, the aircraft would travel 4 feet forward before losing 1 foot of altitude, and 50 feet forward before losing 1 foot of altitude. Needless to say, 50:1 is a better glide ratio than 4:1.

SINK RATE

Equally or even more important than glide ratio, at least when it comes to some aeronautical competitions, is **SINK RATE**. Sink rate is a measure of how much altitude is lost over a standard amount of time. If it's light enough, a plane with a low glide ratio may still have a low sink rate—meaning it may be able to stay aloft longer than a heavier plane with a higher glide ratio. For example, if a plane has a glide ratio of 1:1 but falls at only 1 foot per second, then it will stay in the air longer than a plane with glide ratio of 5:1 that loses 2 feet per second. The key measurement for a duration flight is the number of seconds the plane takes to hit the ground. A good glide ratio can really help, but the lowest sink rate will be the winner, assuming the planes start from the same height.

STALLING

What's a stall? A **STALL** is a loss of lift caused by the airflow becoming chaotic over the wing. This happens when an airplane travels at too slow a speed, when the airplane's wings are slanted too high against the airflow (known as a high **ANGLE OF ATTACK**), or a combination of those two things. When the air can no longer follow the shape of the wing to generate sufficient lift, a stall is the result. (In Figure 10, we see what happens to the airflow as the angle of attack increases—eventually we'll lose lift altogether.)

CONTROL SURFACES

Balancing the four forces we discussed at the beginning of this chapter is the first part of making a good flying machine. Causing or correcting maneuvers is the next step. This is where **CONTROL SURFACES** come into play. *Control surface* is the term used to describe the moving part of any flying surface, which on an airplane includes the rudder, elevator, and ailerons. Elevators control the up and down movement of a plane, rudders control the right and left movement, and ailerons control rolling. All of these control surfaces will be discussed in the sections that follow.

In the name of aerodynamic research only, and never just for fun, put your hand out the window of a moving car. Do not extend your reach past the side-view mirrors. Position your hand horizontally (so, flat and parallel to the ground) with your thumb facing the direction the car is moving. The smallest move of your hand can force it up, down, left, or right. Air bounces off the flat side of your hand that's facing the direction of travel. The deflecting air forces your hand in the opposite direction. If you twist your wrist to point your thumb downward, air is now hitting the back of your hand and bouncing upward. The air bounces up, and your hand is pushed down. Aircraft control surfaces follow this same principle.

Air bounces against them, forcing the plane to maneuver left, right, up, down, or to roll.

Flaps and ailerons are found at the rear of the main wing on conventional airplanes. But for paper airplanes, the elevator and rudder are the most important control surfaces. Technically, most paper airplanes employ a **BLENDED WING**, meaning there's no separation between the main wing and the tail. On blended wings, some control functions are combined; for example, the elevator also acts as the aileron, and is called an **ELEVON**. Don't sweat that. A paper airplane has few moving parts and therefore fewer controls. This makes it a great way to learn how control surfaces work.

RUDDERS

Let's look at a rudder turn. The pilot operates controls that cause the rudder to start deflecting air. In Figure 11, the rudder has been caused to stick out to the right side of the tail. Air will hit that rudder and get deflected to the right. That will push the tail to the left.

Here's the key: The aircraft will rotate around the center of lift in flight. If the tail goes left, the nose will go right. It's like a seesaw, with the center of lift in the middle. The left rudder pushes the nose left because the air gets deflected left, which pushes the tail to the right.

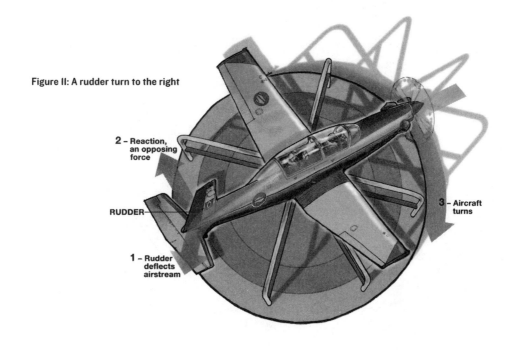

Figure 11: A rudder turn to the right

2 – Reaction, an opposing force

RUDDER

1 – Rudder deflects airstream

3 – Aircraft turns

ELEVATORS

An elevator works exactly the same way; however, elevators control the up-and-down movement of the plane (whereas the rudder controls lateral, or right and left movement).

Controlling the movement of the rudder and elevator is the very basic way to control an airplane. It's all you'll need to get a paper airplane to do what you want. Full-sized planes need more.

Take a look at Figure 12, below. If the pilot raises the elevator, then air will hit the elevator and get bounced up, pushing the tail down and the nose up.

AILERONS

Rudder turns tend to be slow. The plane slips to the side and kind of skids through the air in a wide circle. Full-sized aircraft and really good remote control aircraft use another trick. It's sort of like leaning a bicycle over to make a faster turn. Aircraft lean over too. That's called **ROLLING**. Rolling is controlled with ailerons, shown in Figure 13, below. Ailerons look a lot like elevators, only they're attached to the main wing. The control surface furthest from the body of the plane is usually the aileron on a full-sized plane.

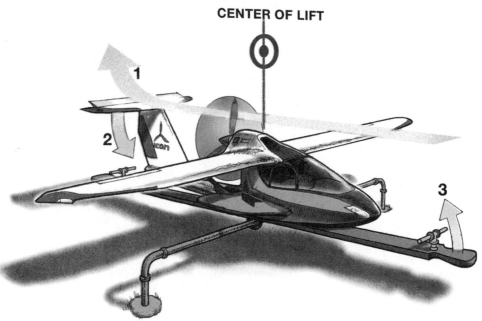

Figure 12: Using up elevator

Figure 13: Using ailerons to roll a plane

Figure 14: Using flaps to increase the lifting surface

Imagine deflecting a little air downward off the right wing only. That wing gets pushed upward. The center of lift seesaw is still working. The left wing dips as the right wing lifts, and the plane leans left. If we add up elevator, the plane will now "climb" in a circle to the left.

You can feel this method of banked turning when taking off or landing at most airports. A basic rudder adjustment doesn't offer the precision and control of an aileron turn. Watch soaring birds like hawks and seagulls execute turns. They use expert aileron-like adjustments of their feathers to make elegantly banked turns.

FLAPS

In addition to ailerons, full-sized aircraft also have **FLAPS**, shown in Figure 14. Flaps have a very basic function: they make the wing bigger during slower flight. Flaps are extended during takeoff and landing to make the wing bigger, adding **LIFTING SURFACE**. Aircraft designers make wings just big enough to lift the plane and cargo at cruising speed. Pushing a bigger wing through the air would only create drag and cost more jet fuel to fly. The problem becomes how to take off and land efficiently. A bigger wing allows the plane to get airborne at a slower speed. Imagine how big an airport would need to be if jumbo jets didn't use flaps. Runways would need to be much longer. The jet would need beefed-up landing gear to handle nearly 300 mph of ground speed before takeoff!

Flaps neatly solve these problems by adding wing surface during slower flight. They get neatly tucked away once the airplane is up to cruising speed. Flaps are usually the control surfaces you see closest to the fuselage. I've yet to invent a paper airplane that has flaps. Perhaps you will, now that you know what they are. It's an intriguing possibility; to have the paper airplane somehow increase wing area during flight.

NOW IT GETS A LITTLE WEIRD

Everyone likes the basic forces diagram we started this chapter with. However, things start to go off the rails when we look at cause and effect for lift. This is the part of flight I really like. There's still room to put your two cents in. It's far from the settled science we're lead to believe it is.

The first thing you need to know is that all of science is really our best hunch about how things work. Pick any point in history and you'll find that about 90 percent of what was "known" at that time about how the universe worked was substantially incorrect. Here's another astounding thing. What we know is usually provable, depending on our ability to observe the thing we're studying. In other words, we're always limited by the tools we can bring to bear. We can't really understand germs and bacteria without a microscope. You can see the effects and come up with theories, but that's not quite the same. Our ability to probe the secret of flight is hampered by the nature of air and gravity. We can constantly see their effects, but they're generally invisible.

For your amusement and contemplation, what follows are a couple of theories regarding the creation of lift—one of the most basic tenets of flight theory. Don't choke on the chalk dust.

BERNOULLI

Most physics teachers explaining flight will begin with Daniel Bernoulli, a Swiss mathematician and renowned theorist of his day (1700–1782), who pioneered work in harmonic vibration, kinetic gases, and, of course, fluid dynamics. I like to mention that he was hired by some monks to get water out of some deep wells. Understanding pressure differentials inside of pipes was critical to the task.

Bernoulli's work with fluids was eventually adopted as the correct way to conceive of and predict airflow. His body of work was completed nearly 180 years before the Wright brothers flew. It took even longer for somebody to connect Bernoulli's work and flight since the discoveries were separated by time, distance, and function. Ah, science . . .

My brother Jim, who just received an electrical engineering degree, was taught the following: Consider the air to be like a slab of jelly. An aircraft wing will slice through the jelly, which will part as the leading edge of the wing penetrates. The jelly on top and bottom get deflected by the surface of the wing they touch.

A wing that's curved more on the top will cause the upper jelly mass to move a greater distance as it follows the curve. From Bernoulli's work, we know that a faster-moving fluid in a closed environment causes a lower pressure. So, naturally, the pressure on top of the wing is lower. What do we call that lower pressure? Lift. The causal relationship is usually stated something like this: the faster-moving jelly (air) causes a lower pressure. If this all sounds confusing, take a look at Figure 15, which illustrates this concept.

Then the professor whips out all the differential calculus to quantify the jelly, the wing surface, and the speed. It's all integrated into Bernoulli's equation. The professor takes a bow, and as the chalk dust settles, you believe another mystery has been solved through the wonders of math and science. Uh, not quite.

Let's review. Bernoulli was working with a closed system: water in pipes. The sky doesn't feel all that closed to me. Some argue that the static pressure from the atmosphere creates a closed system. Perhaps, but that would beg the question "What constitutes an open system?"

Next, the idea implicit in the jelly metaphor is that the jelly on top of the wing gets to the trailing edge of the wing at the same time as the jelly on the bottom. It's an idea referred to as the **EQUAL TRANSIT THEORY**. That ain't true either.

If you color the air streams differently and send bursts of colored air over and under the wing, you can see the air on top gets to the trailing edge first! It's definitely moving faster than the air on the bottom, so the equal transit notion is twaddle. Oops! Let me say, at this point, it's much easier to poke holes in theories than to stitch together a really good one.

A "BUNCH" OF LIFT?

Some aerodynamicists will argue another idea convincingly. I like this one slightly better. The idea is that air kind of bunches up when it encounters the leading edge of the wing. That makes sense. There's friction. Air gets slowed from the imperfections in the wing surface. Add the idea of a more curved upper surface, and now the air is bunching

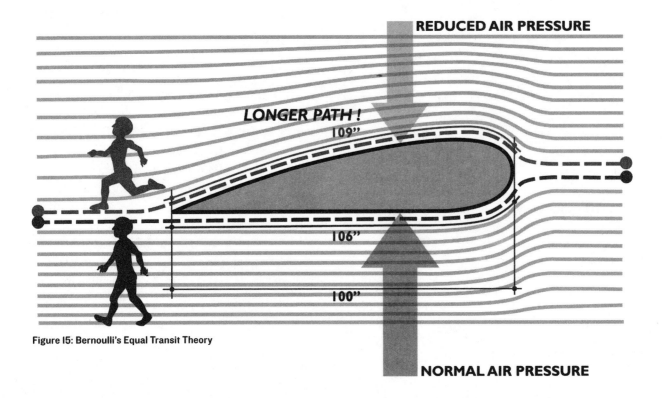

REDUCED AIR PRESSURE

LONGER PATH !

109"

106"

100"

NORMAL AIR PRESSURE

Figure 15: Bernoulli's Equal Transit Theory

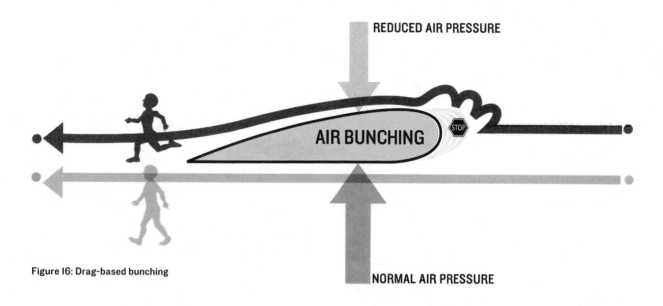

REDUCED AIR PRESSURE

AIR BUNCHING

STOP

Figure 16: Drag-based bunching

NORMAL AIR PRESSURE

up more on the top of the wing. Figure 16 illustrates this bunching effect.

The bunching of the air causes a pressure difference further along the wing. The pressure is lower because the air is bunched up, or under higher pressure, just upstream. Now the air speeds up to fill the lower pressure and scoots off the trailing edge of the wing, ahead of the air on the bottom. The causal relationship is usually stated something like this: the lower pressure causes the faster moving stream of air. Note that this is exactly the opposite of the Bernoulli cause-and-effect statement.

EQUAL AND OPPOSITE ABSTRACTIONS!

Well, now it's getting interesting. We have two plausible but completely conflicting explanations. I love this about flight. This theory of air bunching up at the leading edge seems to fit the facts better. It's not married to a closed system. The air is free to move as fast as we can observe. The pressure systems behave more like most pressure systems we normally hear about: high moves to low. What's not to like?

There's one small problem. A lot of aerobatic aircraft and fighter jets have wings that are curved

equally on both sides. They still manage to generate lift. How do we explain that? A truly acceptable explanation would have to include right-side-up, upside-down, high-speed, and low-speed flight. Are we there yet? Not quite. Another group of flight-minded folks say that it's all about air deflection.

NEWTON? NEWTON? ISAAC NEWTON?

Remember sticking your hand out the window of the moving car? Angled upward, air deflected off the bottom, much like a kite in the wind. This could be how aerobatic planes achieve a good deal of lift. Lots of power makes up for symmetrical wing design, which would yield no lift when parallel to air flow.

Remember back a paragraph or two when we proved that if a wing is curved on top but flat on the bottom, then the air on top is going faster than the air on the bottom? That causes the air at the trailing edge of the wing to shoot downward as it leaves the wing. Some claim flight is just a straight, Newtonian opposite-and-equal-reaction proposition. Air gets thrown down; so the wings get pushed up. Figure 17 on the next page shows the Newtonian theory in action.

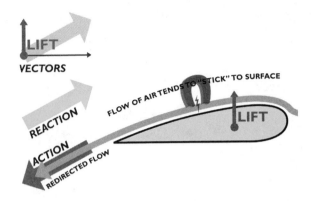

Figure 17: The Newtonian theory of equal and opposite reactions: At the trailing edge, air following the shape of the wing is pushed downward. The wing is therefore pushed upward.

The lower pressure system is just a mechanism that causes the air to get tossed around. Add that to the angle of attack effect, and maybe you have a theory. I like the simplicity. But it doesn't quite account for all that measurable low pressure on the top of wings that have a curved upper surface.

WHAT ABOUT COANDĂ?

Strictly in the name of science, borrow or purchase a beach ball—one of those big, floaty, plastic-pool-toy, inflatable balls. Your next science experiment begins now. Predict what will happen if you throw the ball forward with a lot of backspin (the bottom of the ball is moving the same direction as the ball is traveling). Don't be afraid. I got this wrong on the first try.

First, a little background. The **COANDĂ EFFECT** is the tendency for air (or a fluid) to follow the surface that it contacts. The classic demonstration is water being deflected off-course by the bottom of a spoon. If you've never played with a spoon in stream of water like this, please do so now or take a look at Figure 18. It's possible to dramatically alter the path of the water with the bottom surface of the spoon. Go ahead. I'll wait.

Water or air following the surface of the spoon, wing, or beach ball may be the key to understanding lift.

Did you make a guess about the beach ball? My first guess, which is wrong, was that the ball would curve to the floor. I reasoned that the bottom

of the ball was moving faster through the air, so, like Bernoulli said, faster air means lower pressure and that ball should move that way. *Wrong answer.*

Correct answer: The airflow is trying to *stick* to the surface of the ball. This is all a bit complicated, so take a look at Figure 19 on the next page. Tiny gobs of air stick to the surface and collide with the oncoming air, slowing the flow on the bottom of the ball. The top of the ball is moving the same direction as the airflow, helping the air go faster and hold onto the surface longer. The net effect is lower pressure on the top and air deflected downward. The ball curves upward! This is how a curve ball works in baseball. The ball moves away from the side of the ball moving toward the target. Golf balls curve for the same reason. I wouldn't explain this to an adult during a series of hooking golf shots, though.

Obviously flight is complicated, and I said at the beginning of this chapter, we're not sure why stuff flies. It's likely a mix of all the ideas here and elsewhere. This is a classic case of looking at the facts at hand, developing a working theory, and testing. The idea that the facts should fit into multiple theories shouldn't bother people familiar with the scientific method. This is how science works. Let the best theory win!

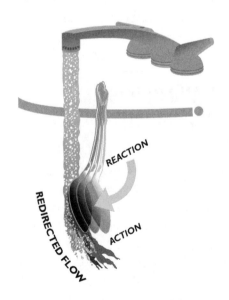

Figure 18: The Coandă effect in action: All along the wing surface, air tends to follow the wing's shape. This is called the Coandă effect. Using a spoon and a slow stream of tap water, you can use the Coandă effect to redirect the stream. Notice that we're using the back (convex) side of the spoon.

THE WILD WEST OF FLIGHT THEORY

The cause-and-effect theories are far from settled science. This is especially true for really small wings like paper airplanes. Through rigorous observation and measurements, we have very good mathematical models that predict with stunning accuracy how a large wing of a certain shape and curve will behave—how much it will lift and when it will stall. Some will argue this is all that's required. They say the rest is just an endless chicken-and-egg argument. In my opinion, that's giving up. Where's the scientific inquiry, I ask you?

What's really great about paper airplanes is that as wings get smaller, the airflow gets more chaotic. The models don't work so well at predicting outcomes. So, we're way out at the western most outpost of the Wild West of flight theory.

What we can observe, document, and prove may eventually lead to a greater understanding of full-scale flight. This may be useful information to memorize if you insist on trying out your newest planes in a classroom. It's a bit of a fudge, but it may get you off the hook if a plane goes wildly off course and bonks the blackboard, or worse, the teacher.

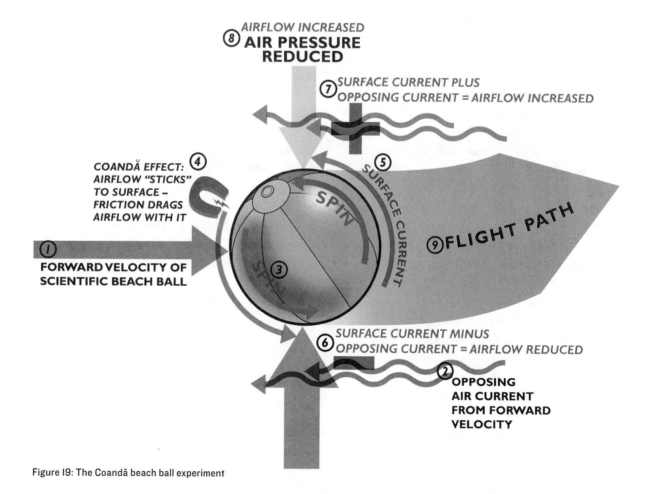

Figure 19: The Coandă beach ball experiment

THROWING AND ADJUSTING

2

Most people have conflicting ideas about paper airplanes. They're sure they used to be able to make a great one, and they're sure they can't make a good one now. It's possible that both of these things are true, but it's more likely that neither one is. What passes for a great flight when you're younger doesn't cut the mustard now, and when you were younger, you were more willing to have a bunch of bad ones to get one good flight. Most people *can* fold a very good plane. And here's where the two ideas—I used to make a good plane and I can't make a good one now—intersect: most people are not very good at throwing and adjusting their planes. Be calm. You've turned to the right page.

What I'm about to tell you works for every airplane—from the most basic, all the way through my world-record plane. To get the full benefit of this chapter, go get one of your paper airplanes now. Or, if you haven't made one, spend some time making one—maybe try one of the designs in this book.

Now I'm going to assume you did a good job folding. The creases are crisp, the wings match, and if you're working from one of the designs in chapter 4 of this book, your plane matches the photo of the finished plane in terms of dihedral, winglet angles, and so on. First, go have some fun. You've earned it. Go ahead. Take a few throws, as hard or as easy as you like. I'll wait.

Did you watch closely what your plane did? Did it do the same thing each time? The more you fly paper airplanes and the more closely you observe them, the more you'll learn about throwing and adjusting. To help you with that flying and observing, throwing and adjusting, I want to share a few tips.

Let's start from the top. Never, ever pick your paper airplane up by the trailing edge of the wing (the tail). I'll show you why. Hold your plane in the palm of your hand. Lightly pinch a rear corner of one wing between your thumb and forefinger. Remove the support hand. See the paper bend? That's really, really bad for the aerodynamics. All the careful adjustments we're about to make will be undone by picking your plane up this way. Okay, put your injured plane down on the table. Smooth the wing out, and look for the thickest set of layers near the nose. There's your handle. Now you're ready to throw.

THROWING A PLANE

Good throws start with a good grip. It's always helpful to hold the plane where the bulk of the layers come together. Generally that's close to the center of gravity—or *CG* as it is known among paper airplane aficionados—on the paper airplane. But where, precisely, is the CG? If you really want to know, it's not that hard to find out. Get a needle and thread. Start running your plane through with the needle and let it hang from the thread. Where the plane balances with wings perfectly flat, hanging from the thread, there's your CG.

Back to the proper grip technique: a 1980s hair band, .38 Special, had song lyrics that fit here—"hold on loosely, don't let go."

That's the kind of pressure you need to apply for a good grip, firm enough to keep holding but not so tight you'll bunch up layers or warp the wings. Or as .38 Special put it, "If you cling too tightly . . . you're gonna lose control." I hope that was funny to someone.

Good grip

Too tight: Layers are bunching up on the left wing.

MAKING THROWING ADJUSTMENTS

Once you know how to throw a plane, you'll also need to learn how to make adjustments to get the plane to do what you want it to do. To make the proper adjustments, the first thing you do is observe carefully what the plane does after you throw it.

VEERING RIGHT OR LEFT

Is the plane veering one direction or the other just after launch, but it straightens out later? Could be you need a thumb adjustment. I'm going to assume you're right-handed for a moment. If not, reverse the following directions. If your plane veers right, move your thumb a little lower on your grip. (Lefties go higher.) If you just hold your plane and slide your thumb up and down, you'll see what we're up to here—take a look at the photos on the next page. Changing your thumb position actually changes the angle at which the plane is released. Just a little thumb movement will affect the launch angle a lot. Check it out: Rock the plane left or right by sliding your gripping thumb. (Joe, my thrower, figured this one out for our world-record launch. That's why he's a quarterback and I make planes.) If the plane veers left before flying straight, try raising the thumb position a bit. (Lefties go lower.)

STALLING

Is the plane just climbing up, stalling, and falling? Try releasing the plane at a much lower angle. I always start with a flat toss, straight out at shoulder height. I try to assist the plane into the air for the first few throws. After making a few planes, you'll pick up a knack for guessing about how fast a plane is going to fly. Try to release the craft at that speed. Help it into a stable glide, and then see what needs fixing. Something always needs fixing. Remember, it's a flying machine and made from paper. The moment you get it folded, the paper is trying to unfold. It might be drying out or soaking up moisture. It will change shape over time. To keep it flying right, you need to keep adjusting.

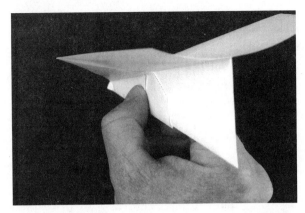

Midposition on the thumb

Thumb is lower, rocking the plane over. This is exaggerated to demonstrate the idea.

Thumb is high, rocking the plane the other way.

MAKING CONTROL SURFACE ADJUSTMENTS

Changing your throwing technique is one way to correct your plane's flight. However, you can also change the plane itself by adjusting its various control surfaces.

HOW DO I MAKE MY PLANE TURN RIGHT OR LEFT?

If you want to change the lateral direction of your plane, you should make some rudder adjustments.

Just where the wings meet the body of your plane is the best spot to play with the rudder. Sometimes you can get **WINGLETS**, which are small, vertical fins on the wingtips, to work well for a rudder adjustment, especially if you've already adjusted where the wing meets the body.

If you hold your plane as in the picture below, then bending the vertical part of the fuselage to the right will make the plane go right. It's my job to bore you with the aerodynamic details, so here goes: The air travels down the side of the plane and hits that bend. The air will get deflected to the right, and that pushes the tail to the left. Since the aircraft rotates around the center of lift in flight, the nose of the plane will pivot right when the tail is pushed to the left.

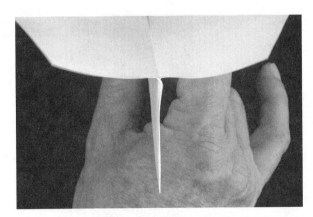

A lot of right rudder adjustment, to demonstrate the idea. Go much smaller to start.

Don't bother stuffing all those technical details between your ears. Just remember to bend left to go left and bend right to go right. Start with very small bends, half a finger-width long and just a millimeter or so wide. Just that much will make a huge difference.

HOW DO I MAKE MY PLANE CLIMB OR DIVE?

You may have guessed by now, but I'll tell you anyway. The same logic works for up and down adjustments. Where the wings meet the body of the plane, you bend upward to go up and downward to go down. Take a look at the picture below, where I've added some up elevator. The same caution with bending too much applies here. A little goes a long way.

Up elevator adjustment as seen from the top of the plane

HOW DO I JUST MAKE THIS PLANE FLY STRAIGHT?

"Okay," you say, "but I want my plane to fly straight. I'm entered in a distance contest. I don't want it turning." Ah ha! You've come to the right place. If your plane is turning left, bend that vertical surface a little to the right. Try a small adjustment and then throw the plane. Did it help? By making the plane turn against its bad behavior, you can get it flying straight. This strategy works for too much left, right, up, or down trajectory. Adjust against the offending direction of travel.

I FOLDED THIS PERFECT AND IT STILL TURNS. WHAT'S UP WITH THAT?

Why does my plane turn one direction or another, even when it looks perfectly well folded? It could be tiny imperfections in one crease or another. It could be a slight difference in dihedral angles of the wings. It might be that the layers on one wing bulge slightly more than the layers on the other wing. Any one of these things can be hard to see and even harder to correct. It's much easier to correct for the overall effect of all the flaws than try to address each and every one individually. Trust me. I've tried.

I could tell you that adjusting is an art, but I'm not sure it's all that. Most people get the basics pretty quickly, but most people also tend to overadjust. How do you deal with that? It's a little like playing Wii. Doing it a bunch makes you better.

Adjusting is a series of compromises. Every time you correct a bad thing, you add another bad thing: drag. Every bend you make slows down the plane, making it harder to slip neatly through the air. I offer this as food for thought—and a really good reason to fold carefully and keep your adjustments small.

FOLDING 101: THE BASICS

3

First and foremost, you should refer to the folding symbols on page 128, which explain how to use the step-by-step photos for all the planes in this book. Apart from that, there are a few all-purpose techniques worth knowing when it comes to paper folding. The best way to fold a sheet in half is one. Making a great diagonal fold is another. I've watched so many people do it wrong that I've finally decided it's worth putting the correct method in a book.

First things first: When folding, choose a flat, clean, and dry surface. I like glass, but hard plastic is even better. Find the smoothest surface you can. (I actually had a piece of hard plastic made by a local plastic fabricator. The world record plane was folded on 1/4-inch thick, polished plastic.)

Now, let's start with the move you'll do over and over throughout the book: folding a sheet of paper in half.

Most people start at one end with a couple of corners lined up and sweep their way to the opposite end; controlling the other corners seems to be optional. They're all about getting a straight crease at all costs including misaligned corners at the far end. The misaligned, whopper-jawed end always becomes the tail because you think all the layered, accurate folding needs to be in the nose, right? If you only change one thing about your folding, let this be it. Use the messed-up end for the nose. Matching corners for the tail, please.

HOW TO FOLD A PAGE IN HALF (YEAH, THERE'S A RIGHT WAY)

The better method is to line up all the corners. Keep them pressed together with your index fingers while you find the center of the new crease with your thumbs. Press the center flat and sweep outward, toward the ends. You'll get half the error you would if you sweep from one end to the other, and both ends will line up properly. Choose the most perfect one for the tail.

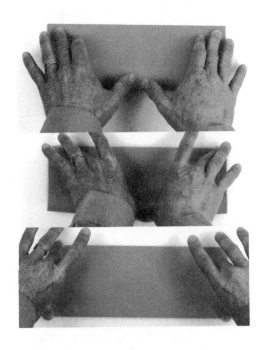

HOW TO MAKE A DIAGONAL FOLD

Here's another one people butcher all the time. They frequently tend to scrunch the layers at the last second to make it "come out right." Try the technique illustrated below instead, which requires carefully splitting a corner in half.

DIAGONAL FOLD

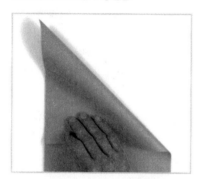

1. Choose the corner you'll be folding through. Move the top of the paper over to the side with which it will line up when the crease is made. Start pressing the layers together, but don't make a crease.

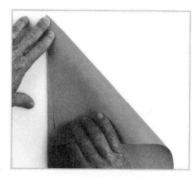

2. With the top and the side lined up roughly, slide your fingers up to the corner to be creased. Start the fold by pinching that corner between your finger and the tabletop. By pinning the corner down, you'll be able to fine-tune the positioning of the layers before making the crease.

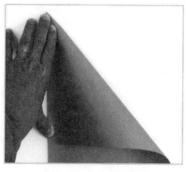

3. Use the pinched corner as a pivot point to swing the top very accurately against the side.

4. Once it's lined up, just sweep the crease into place by starting at the pinched corner and moving down and away.

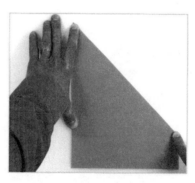

5. Take care to keep the corner pinned down and the edges lined up.

CLASSIC ORIGAMI FOLDS

MOUNTAINS AND VALLEYS

All folding boils down to this: mountain folds and valley folds. When viewed from above, the result of your fold, the crease, will either point up at you, or point down and away from you.

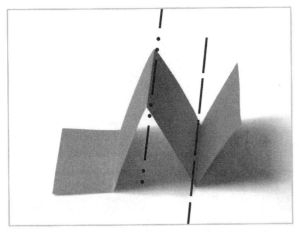

A crease that sticks up like a mountain has been **MOUNTAIN FOLDED**. A crease that looks like a little valley (maybe to an ant) has been **VALLEY FOLDED**.

THE WATERBOMB BASE

The photo sequence on page 27 illustrates how to create a waterbomb base, step by step. All it takes is just three creases and this amazing base happens.

I've turned the top flaps into landing gear, tunnels, canards, and more. Where you take this base is up to you. Go for it. This is ground zero for my paper airplane–making career. It's so simple and yet deliciously complex at the same time.

THE SINK FOLD

Usually some wise guy like me will come up with a reason to poke part of the pointy end of a waterbomb base down inside itself. When you do that, it's called a "sink fold." I suppose it might look like the top of a mountain that's just sunk into itself. It's more likely that rookies get that "sinking feeling" when they see the instruction to sink fold. The

key is making a square inside the big square and then making a little replica of the bigger waterbomb base. The difference? The small one has all the creases going the opposite direction from the big one. The photo sequence on page 28 illustrates this fold.

THE REVERSE FOLD

Think of a long, pointy shape that's been creased in half the long way. A portion of the long crease is getting the direction reversed from a valley to a mountain. Everything around the point of reversal has to compensate. The crease is such a diva. There are two possible ways to reverse a crease, to the outside and to the inside. In this book, I've used both types of reverse folds. The photo sequence on page 29 illustrates an inside and an outside reverse fold.

THE SQUASH FOLD

This fold, illustrated on page 30, is one of the more satisfying moves in paper airplane making. You can move multiple layers by opening them up and squashing the whole mess flat. Generally, if you keep the center creases lined up, this is one of the easiest techniques to get right.

THE PETAL FOLD

This is serious origami. If you're not interested in creating appendages like landing gear, or bird heads, you don't need this technique. It's a lot of folding, but it's the only way to make certain things hold together. There are three preliminary creases: you'll find the petal fold illustrated on page 31.

• • •

It's worth noting here that the world-record distance plane contains no fancy folding at all. Suzanne has just eight creases. When properly made, the creases allow the plane to travel further than any paper airplane in history. Getting the simple things right matters immensely.

THE WATERBOMB BASE

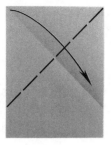

I. Make a couple of diagonal folds.

2. Flip the paper over.

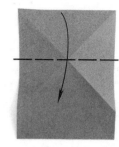

3. Make a crease that cuts across the intersection.

4. Leave that creased and flip the paper over.

5. Open up the crease enough for the paper to stand up.

6. Press down where all the creases meet.

7. Keep pressing . . .

8. . . . until the paper "pops" the other direction.

9. Bring the top down as the sides come in.

10. Some people find it easier to hold the bottom layers down as they press the top layer into position.

11. To finish the fold, press the top layer flat.

12. The completed waterbomb base. I still use this in many planes.

THE SINK FOLD

  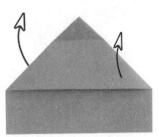

1. Start with a waterbomb, then make a crease at the place you need to sink.

2. Unfold that crease.

3. Open the whole page up and flip it over so that the diagonal folds are mountains.

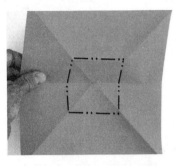  

4. Turn that little square into a mountain folded square; go all the way around it with mountain folds.

5. Now make all the creases inside the little square go in the opposite direction from what they do now. If it's a valley, make it a mountain. If it's a mountain, make it a valley.

6. Remake your original waterbomb while holding the little mountain folds around the square in place. The whole thing should now collapse together.

 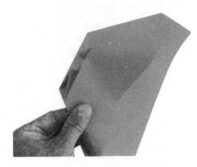 

7. From the side, you can see the sunken part looks like a small waterbomb base.

8. Start pressing the layers together . . .

9. . . . and the original waterbomb layers should line up nicely.

10. The finished sink fold. It'll look like the top sunk, sort of. I admit, it seems like a wacky way to move layers around, but sometimes there's not a better way to get to your goal.

THE REVERSE FOLD

1.   This is a typical starting point for a reverse fold: a narrow point of paper with a long creased edge on one side and open layers on the other side.

2.   Start by pulling the point down.

3.   Pinch together the sides of the pulled-down part to form a valley fold down the middle. Notice that even this early, the outside part is now inside the pinched part.

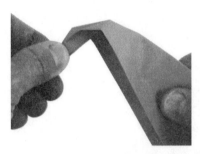

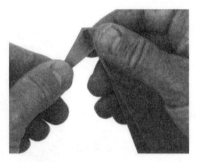

4.   Holding the pinched-together part, position the point to the correct location.

5.   Flatten the fold to lock the reversed layers in place.

6.   The finished reverse fold.

ALTERNATE REVERSE FOLD METHOD

1.   There's an alternate way to begin, which is to fold the point down to make an initial crease where the reverse fold will ultimately be positioned.

2.   Unfold, then make a reverse fold by reversing the direction of the initial crease.

3.   Here's the finished inside reverse.

4.   It's possible to move the inside layers to the outside to create an outside reverse fold (pictured on the left). Compare that to the inside reverse fold (pictured on the right).

THE SQUASH FOLD

1. Start with a waterbomb base; this will give you four flaps on which to practice your squash fold technique. To make the squash fold, move the right point of the top layer all the way to the top tip of the waterbomb base. Make a crease.

2. Unfold the crease about halfway.

3. Open up the layers.

4. Start pressing downward on the crease that joins the opened layers. As the crease touches the bottom layers, take care to line it up with the creases below.

5. Press the flap flat and crease well. Here's your result: a squash fold.

THE PETAL FOLD

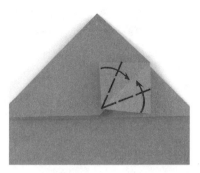

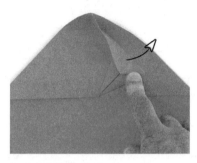

1. Start with a squash fold, then fold the outside edges to the center to create a kite shape.

2. The top of the kite gets folded down.

3. Unfold the flap . . .

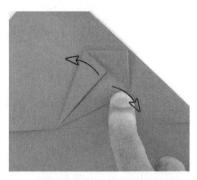

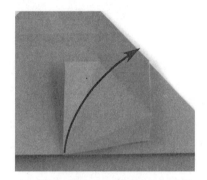

4. . . . then unfold the remaining two flaps.

5. Notice the triangle of creases. On the top layer, only the short crease gets to stay a valley fold. Lift up this top layer, using the short valley fold as a hinge.

6. Here is the top layer about halfway opened up.

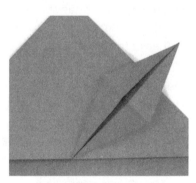

7. Allow the bottom layer to use the valley folds as you move the raw edges to the middle.

8. You'll need to switch the direction of the creases on the top half to get the raw edges to the center. Once all the raw edges meet up at the center, press the whole thing flat.

9. The finished petal fold.

THE PLANES

4

This chapter kicks off with the world record plane, Suzanne. I know you'll give folding that one a try. Afterwards, I hope you'll venture further into the world of folded flight. I'm very proud to present a collection of planes and follow foils (see page 87) that would comprise a great paper airplane book on their own, without the bonus of a world record plane. I've been waiting almost 10 years to share some of these designs in book form. I hope you'll fold some great flying planes and catch the bug to start inventing your own.

If you do decide to start with Suzanne, you will need to gather some supplies. (The rest of the planes in this book require only a sheet of standard 8½ by 11-inch paper.) Below is a list of the equipment and materials I used to create the world record–breaking plane:

1. A bone folder to make sharp creases. You can find them on the Web by searching for "bone folder." Other things can be used for this purpose. The folder just needs to be perfectly smooth, small enough to make creases, and large enough to hang on to. I designed one that I think works pretty well. To purchase my design, go to my website, www.thepaperairplaneguy.com.

2. A snack clip to help hold the plane together during taping. I prefer the ones with little rubber pads inside. They are gentler on the paper.

3. Light-duty cellophane tape, 25 mm wide. Depending on where you live, this is also called sticky tape or Scotch tape. Guinness rules don't require you to use tape; they simply allow a 25 mm by 30 mm piece.

4. A metric ruler. I use this mainly to create my measuring device (see step 19 on page 35), which measures the correct length of tape (30 mm) and my mid-wing tape placement.

5. A pair of very sharp scissors.

6. A hobby blade or other small implement for transferring tape strips. A plastic chopstick would work. This doesn't have to be sharp.

7. A protractor. This will be used to make dihedral angle gauges in the final steps of adjusting the plane. Any brand will work. If you don't have one handy, www.barryscientific.com was just granted a patent for a new design that's probably the most accurate one on the planet.

8. For a world-record plane, I recommend Conqueror CX22, Diamond White, 100 gsm (grams per square meter), unwatermarked, A4 paper. For me, my best throws came with 100 gsm laid paper made by Conqueror. The ridges helped my throws, but as described later, they hurt Joe's higher speed throws.

   A4 is difficult to come by in the US. A piece of regular letter size paper (8½ by 11 inches) can be cut into the correct ratio by removing 19 mm from one long side. Also, 26lb stock is a rough approximation of 100 gsm. If you're planning on breaking the record, be prepared to spend money on shipping paper from the UK. If you're lucky enough to live where international paper sizes are in common use, you're ahead of the game.

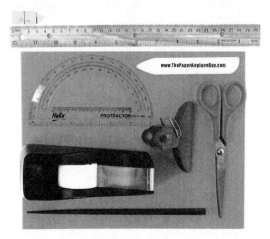

THE WORLD-RECORD PLANE:
SUZANNE

This is the plane Joe Ayoob threw for a Guinness World Record, shattering the old mark by 19½ feet. The world record we set on February 26, 2012, is 226 feet, 10 inches. Joe has thrown this design 240 feet in practice, a full 85 yards. We now find ourselves limited by the size of the hangar. In short, we really don't know how far it will go. Perhaps you will be the person to find the true limit.

A word of caution here: This plane contains only eight creases. Don't let that fool you. A high degree of precision in the folding, taping, and adjusting is required. The taping scheme alone may drive you batty. This is the most technical paper airplane I've ever made. Materials, construction technique, and adjusting for flight require every bit of attention to detail you can muster. Happy folding and flying!

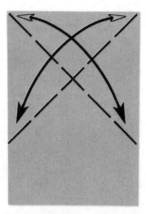

1. Start with the short side up. Fold the top against the left side, making a big diagonal fold that precisely splits the upper left corner.
2. Unfold step 1.
3. Fold and unfold the other diagonal.

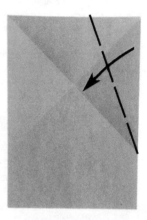

4. Fold the right side of the page against the crease. Leave a little space. How much is a little **space**? See step 5.

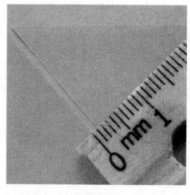

5. Note the gap between the edge and the crease in the photo. It's about a millimeter. This gap should be at least 1 mm and less than 2 mm. Unfold step 4.

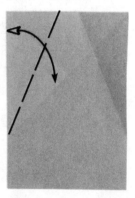

6. Fold and unfold step 4 on the left.

7. Now, remake the creases from steps 4 and 6; the right side first is my preference. Why would I tell you to unfold step 6, and now fold it again? Because you need to do exactly the same thing to both sides of a world record plane to make it fly correctly. Also flatten the horizontal creases I marked, using your bone folder or other smooth object.

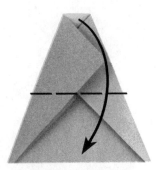

8. Make a valley fold that crosses the center of the diagonal folds.

9. Following the crease, fold the right corner toward the center. Before doing this step, double check that your creases have lined up perfectly from step 8. That will make the folding go easier.

10. Unfold step 9.

11. Following the crease on the left side, fold that corner toward the center.

12. Before remaking the right-side crease, flex the crease from step 11 open and closed once.

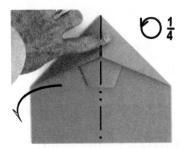

13. Fold the plane in half. This is a mountain fold. Get the nose working first, then line up the rear corners precisely before nailing the crease. I found that, particularly with heavier paper stock like this, making the center crease first is counter-productive. The layers will try to follow that center crease and warp the shape of the plane. It's better in this case to carefully craft the center crease now, making equal amounts of paper drift to each side. Rotate the plane a quarter turn to the left.

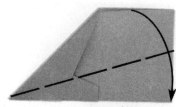

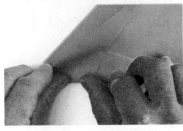

14. This is the Phoenix type of wing fold I started using way back in the 1980s. Start close to the nose and roll the wing over so that the raw edge just touches the rear corner of the plane. Note the bottom photo. My thumb is wedged against the layers along the center crease. It's critical to keep these layers together as you make the wing crease.

15. Start rolling the wing over. The middle photo shows the wing rolling over to meet the rear corner. Line up the edge with the corner exactly. Flip the plane over.

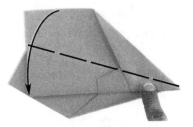

16. Make the wing fold on this side.

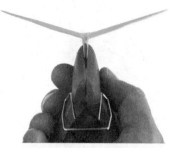

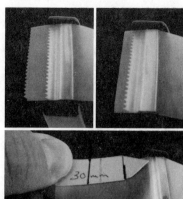

17. Flatten all the creases on both sides with the bone folder. Time to use the snack clip. Spread the wings. Keep all the layers lined up as you apply the clip. The first piece of tape will go on the layers marked by the circle, so keep the clip clear of those layers.

18. This is how Suzanne looks after the wings are made. The folding of the plane is now complete; the rest is taping and adjusting. This is the point where I usually toss a plane out if the folding didn't go well. The first thing to check is the tail. Do the wing creases meet perfectly here? If not, you might as well start over. You won't break any records if this tail surface isn't perfectly flat at the joint. Are the layers under the wings bubbling up the same on both sides? Big errors here will ruin your chances.

19. Tape time. Remove the jagged edge of the end of tape. Don't attempt to use the tape dispenser to cut your tape pieces. Madness lies in that direction. I made a measuring device, which was easier than holding ruler and trying to cut the tape. Of course a ruler could be used, but I found that hopelessly clunky. I like to stick my 30 mm piece of tape to the back of the dispenser before I start cutting off strips.

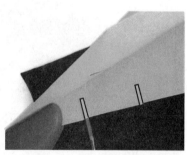

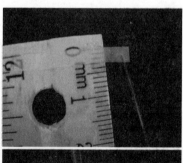

20. The strip in the bottom photo is the standard width for most of the tape pieces. It's about 2 mm wide. I'm holding the tape strip above the ruler in the photo, so the strip is slightly smaller than it appears in the picture. You might want to take a few practice snips before continuing. You'll need a very sharp scissors and a steady hand. From here on out, you'll need to be good at making very accurate 2 mm strips of tape.

The first strip wraps around the center crease of the plane marked by a circle in step 17. Place half on one side and then wrap it around. Use the snack clip to keep the layers in place as you apply the tape as shown in the bottom photo.

21. Reposition the clip more forward to allow room for the next strip. Notice how the tape is across the top layer. This piece is also 2 mm and wraps around the center crease. Keep the layers pressed together evenly and neatly.

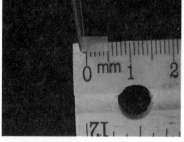

22. The next strip is 3 mm to 4 mm wide. The top photo shows the whole strip. This 4 mm-wide strip will get cut into three pieces. To do this, first cut away a 6 mm piece, as shown in the lower photo. Then, cut the remaining piece in half.

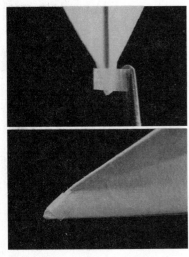

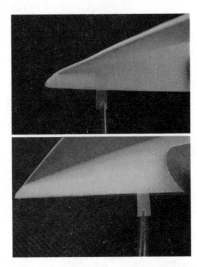

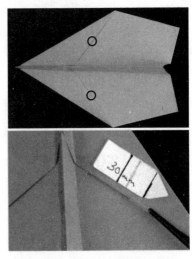

23. The 6 mm piece goes across the top side of the nose. Holding this together helps the structure and airflow. The bottom photo shows the tape applied.

24. After taping the top of the nose, the other two parts of the 4 mm piece get wrapped around the bottom of the nose layers. Keep the layers pressed tightly together as you apply these pieces. The forward piece is 20 to 22 mm from the nose (about I inch). The second piece is applied halfway between the first piece and the piece from step 2I.

25. Cut another 2 mm piece, and then cut that in half. One half of the strip goes on the bottom of each wing. My handy measuring device shows the position: 30 mm from where the top two layers meet near the fuselage. The center of the strips gets placed right there.

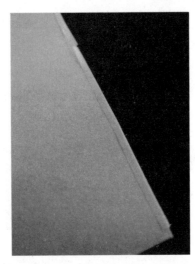

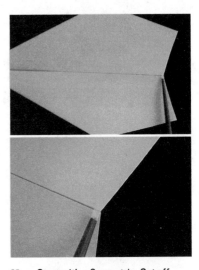

26. Now the tail. This is really hard. You're going to lock the tail shut with a couple of 2 mm strips. One goes on the side closest to the camera. Wrap the strip lengthwise around the tail. The tiny strip of tape is like a hotdog bun wrapped around the tail. The bottom photo shows the positioning of the first strip for the tail.

27. Here is step 26 completed. Flatten both pieces with your bone folder.

28. Start with a 2 mm strip. Cut off 6 mm from one end. The top photo shows the piece on a chopstick, getting put into position. The top side of the tail gets locked together with this tiny piece. This ensures that the top of the tail is like one solid layer of paper. The bottom photo shows the placement. I mentioned this earlier, but it's worth saying again. The trailing edge of the wing needs to be one smooth layer. An error here will cause erratic flights.

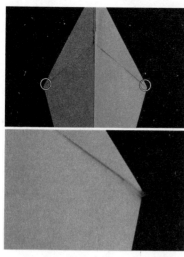

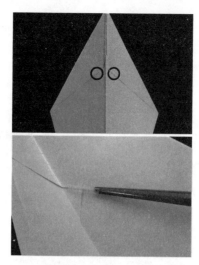

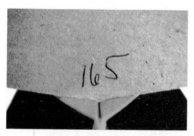

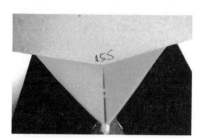

29. Flip the plane over after step 28. Cut the remaining length of tape from step 28 in half. Each half gets placed near the wing tips to hold that layer in place.

30. Cut another 2 mm piece of tape and cut it in half. Each half gets placed as shown. This is the final pair of under-wing pieces.

31. The remaining piece of tape gets cut down the center, generating a couple of relatively wide strips that are used to hold the wings together. Leave a small gap between them. That definitely helps these pieces work better.

32. Time for the final adjustments. This photo shows how your plane should look now. Where the wings meet, the nose gets flattened. I've used my protractor to cut a piece of cardboard (from a cereal box) with a 165-degree interior angle. Dihedral angle actually gets measured from below the wings and from the horizontal—so the dihedral measurement here is 7.5 degrees. This is a much simpler method of achieving that.

33. This photo shows the midwing measurement of 155 degrees. That's the interior angle of another piece of cardboard, generated with the protractor. Again, the actual dihedral, when measured in the standard fashion, is 12.5 degrees.

34. This photo shows the tail dihedral calibrated with our 165 degree chunk of cardboard. This lower angle at the tail helps create less drag.

35. The last step is critical to breaking the world record. Reduce overall drag by flattening all the edges of the tape pieces. Burnish them in place with a bone folder or similar device.

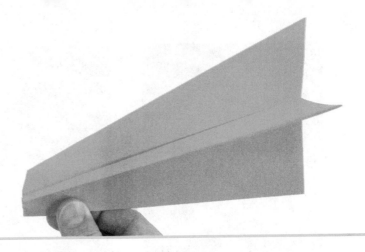

VERY EASY

I wanted a very-easy-to-build, higher-performance design. This is a very clean design, made to compete with the likes of the legendary Nakamura Lock, a paper airplane design that's been a schoolyard standard for decades. The ease of construction will please the beginners, and the great flights will please those with a bit more skill. Practice precision on this one. Make your folds and your creases crisp. With so few steps, you can make a few of these quickly.

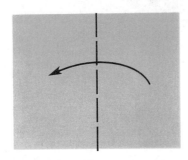

1. Start with the long side up. Fold the page in half.
2. Unfold.

3. Flip the page over.

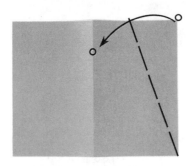

4. Move the marked corner to the center crease. Fold through the lower right corner.

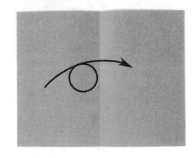

5. Repeat step 4 for the left side.

6. Line up the creased edge with the raw edge, all the way down to the lower left corner.

7. Repeat step 6—this time on the right.

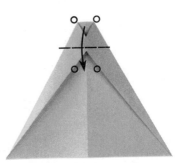

8. Fold the top edge down so the corners meet the marked edges.

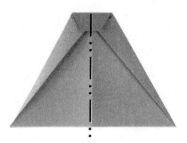

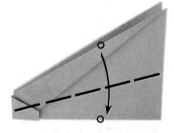

9. Fold the plane in half. Watch out, this is a mountain fold. Flip the plane over, make a valley fold, and then flip it back over. That's how to create a mountain fold.

10. Rotate the plane a quarter turn to the left.

11. Make a very straightforward wing fold: take the top edge of the wing and line it up with the center crease from step 9.

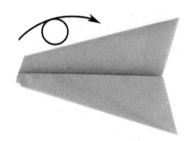

12. Flip the plane over and make the other wing.

13. Make the wings match and your folding work is finished.

14. This is a great place to start learning to throw and adjust. Test-fly yours after making it match the photos.

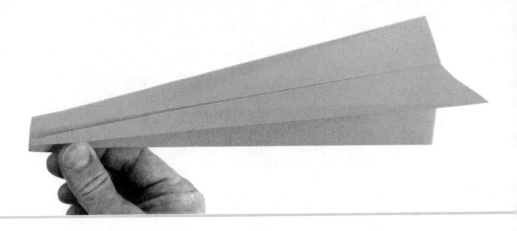

JAVELIN

Very early in the record chase, I developed this plane. It turned out to be a great distance model that could add a little glide action to the ballistic toss. It also turned out to be too much of a dart and not enough of a glider to conquer the record. The ease of folding and tuning make this a natural addition to any great hangar full of paper airplanes.

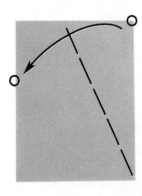

1. Start with the short side on top. The crease starts at the lower right corner and puts the upper right corner against the left edge. There's only one way all of that lines up. Make the crease.

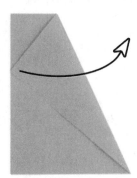

2. Unfold.

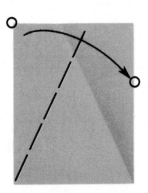

3. It's the same move for the left side: touch the left upper corner to the right side and let the crease hit the lower left corner.

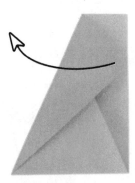

4. Unfold.

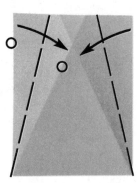

5. Line up the edges with their closest creases and make valley folds.

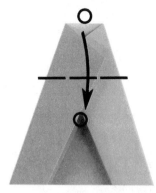

6. Move the creased edges to the raw edges and make valley folds through the bottom corners.

7A. This photo is the finished step 6.

7B. This photo is a closer view, showing you how to move the top edge down to meet the point where the layers cross.

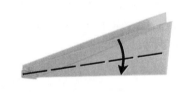

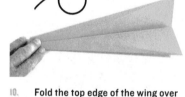

8. The top photo is the finished step 7. The bottom photo asks you to fold the plane in half. Careful, that's a mountain fold there.

9. Rotate the plane a quarter turn to the left before making the wing folds.

10. Fold the top edge of the wing over to meet the center crease. The bottom shows you the result and asks you to flip the plane over to make the other wing crease.

11. Here's that other wing folded down. Now spread those wings and get ready to fly.

12. The finished Javelin. It's got enough wing area to float a bit, and you can see it's built for speed.

THE FLOATER

The Floater is balanced to give you floating, slow flights. With the nearly flat dihedral and the down-turned winglets, this plane has a fun design and great flight potential.

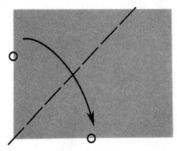

1. Start with the long side on top. Fold one short side to the bottom. This is a diagonal fold, cutting the bottom left corner in half.

2. Unfold.

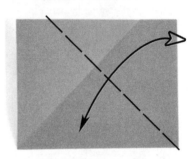

3. Repeat steps 1 and 2 for the right side of the paper. Fold and unfold.

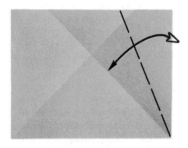

4. Fold one side over and align with the crease, then unfold.

5. Fold the same side over to align with the crease from step 4.

6. Remake the crease from Step 4.

7. Repeat steps 4 through 7 on the left side.

8. Move the top edge down so it meets the points where the original diagonal folds meet the edges marked with circles in the photo. Make the crease.

9. Fold the layered parts in half.

10A. Fold over again, following the creased edge of the layers.

10B. Here is the finished fold.

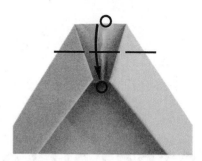

11. The zoomed-in view shows the next step. Fold the top down to meet the marked edge.

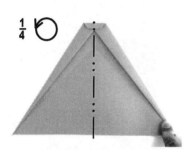

12. Fold the plane in half, keeping all the layers on the outside. You can do this by flipping the plane over and then folding the plane in half. Rotate the plane one quarter turn.

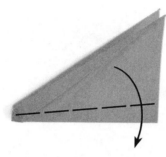

13. Create the wing by making a fold that slopes gently upward toward the tail. Start at the nose, half way up the vertical edge.

14. Flip the plane over and make the other wing. First, a quick peek at the fold you just made. Note the gentle slope upward toward the tail.

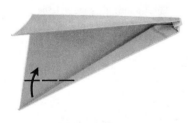

15. Here, on top, is the other wing matched up and finished. The bottom shows the winglet fold. Line up the same angle as the main wing fold (parallel to that) and fold up about one third.

16. Flip the plane over and make the other winglet fold. Note that you can use the first creased winglet edge to line up the second one. This is a handy technique.

17. The finished Floater. Notice the nearly flat dihedral angle. Also check out the up elevator. That should be enough to keep this one floating.

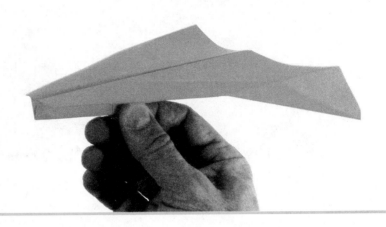

STRETCH LOCK

The lock in this design is akin to the Nakamura Lock layer-controlling system. The open fuselage and downward swept winglets create a fanciful look in flight. It's structurally sound enough for outdoor action.

1. Start with the long side up. Make a diagonal fold from the upper left corner to the lower right corner. This is one of the hardest folds to make accurately. Take your time.

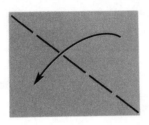

2. Unfold.

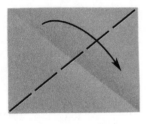

3. Your reward? You get to do the other diagonal.

4. Unfold.

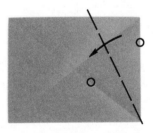

5. Line up the right edge with the diagonal marked and make the valley fold.

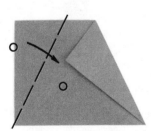

6. Once more, on the left, the side goes to the diagonal crease.

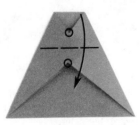

7. This is weird. Make a fold so that the circled points align, then crease. You'll have to lift the raw edge up to peek as you make the crease.

8. The creased edges get folded to line up with the raw edges.

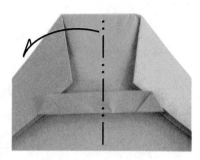

¼

9. Lift up the tab to cover the other layers. Eventually this will lock the plane together.

10. This is a closer view of the completed step 9. Mountain fold the plane in half.

11. Rotate the plane one quarter turn to the left.

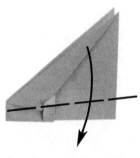

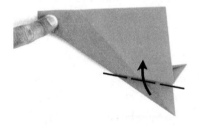

12. Start the crease about half way up the vertical nose edge. End about even with the top of that vertical nose-edge height. It's a gentle upward slope as the crease goes aft.

13A. Here is step 12 complete.

13B. This gives you a look at the finished crease before you turn the plane over and make the other wing by repeating step 12 for the other side.

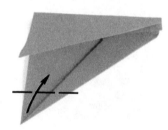

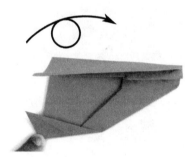

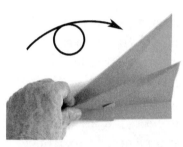

14. With the other wing made, allow the wings to spring open a bit. The winglets will point down on this plane, so we're making valley folds—about one-third of the width of the wing—on the underside of the wing.

15. Flip it over. I want to show you a neat trick for making the winglets match.

16A. Leave the first winglet folded and press the wings together. Fold up and over the first winglet crease to make the matching winglet on this side. The next photo shows the completed move.

16B. Here are the finished winglets.

17. The finished Stretch Lock. Note the flat dihedral angle and the winglets angled slightly out.

PRO GLIDER

If I ever go for the world record for duration, this plane will serve as the starting point for my design. The clean lines, locked-together body, and great balance will give you plenty of fun. This is the plane where I invented the Collins nose lock. I had just seen a great fuselage-locking technique from Takuo Toda, the current world record holder for duration, and was completely jealous. Locking the body of the plane together creates less drag and helps keep the dihedral angle precisely where you adjust it. I'd used a squash fold way back in my book *Fantastic Flight* for the LF-1. Toda used a squash fold for even better effect. It was simple and perfect. I believe this may be aerodynamically cleaner than Toda's locking fold. Try it for yourself— and don't neglect the alternate wing crease. I really like that look.

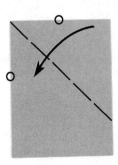

1. Start with the short side up. Make a diagonal fold by lining up the top edge with the left side of the page. Split the left upper corner in half with the crease.

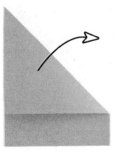

2. Unfold.

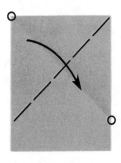

3. Make a diagonal fold the other direction. If you were very accurate with the first step, you'll be able to bring the upper left corner to the other end of the crease to make this step.

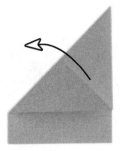

4. Unfold.

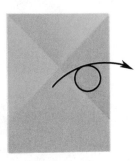

5. Flip the page over.

6. Fold in half the long way.

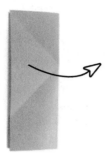

7. Unfold.

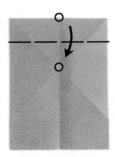

8. Fold the top edge down to meet the point where all the creases intersect. Line up the end of the crease on the top edge with the spot where all the creases meet. Notice in step 9, there's a little space between those points. That's to allow for the creases that are on the way.

9. Fold the layered part in half. Use the ends of the crease to keep the layers lined up.

10. Rotate the plane a quarter turn to the left.

11. Fold the layered part to the right. This crease should hit the point where the diagonal creases meet.

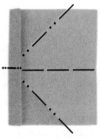

12A. This is a complex move. Study this diagram closely before making your actual folds. Note the little mountain fold on the layered section. The valley fold extends all the way to the left side of the back layer only.

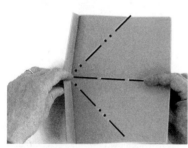

12B. To begin folding, lift up the layered part and start pinching the mountain fold together. Start making the valley fold.

12C. It's all coming together now. Keep pinching the layered part together. Collapse the whole thing flat using the mountain folds and valley fold.

12D. Are you there yet?

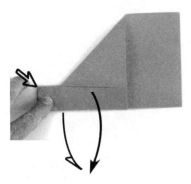

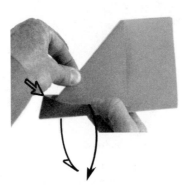

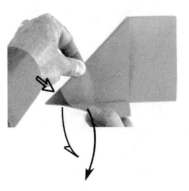

13A. Allow the marked corner to move down as you pull the layered parts out and down. This is hard.

13B. Now actually pull the layered parts on each side of the plane out and down. See the corner start to move down?

13C. Notice the corner is now against the other layers. Keep pulling the outer layers on each side of the plane down. We'll flatten the whole mess in a moment.

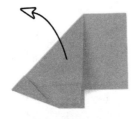

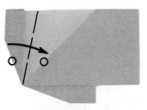

13D. Pinch the layers all flat. Start at the nose and move toward the tail to flatten the layers.

14. You've just completed the heart of a great nose-locking technique! No time to rest. Lift up the top layer to get ready for the next crease.

15. Move the marked edge to line up with marked crease. Leave about a crease's width of space to allow the next fold to go smoothly.

16. Remake the crease.

17. Flip the plane over. We have to do the other side.

18. Remember how this goes? Lift up the layer, make the crease, and then refold the crease (steps 15 and 16).

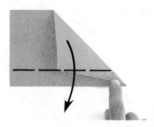

19A. Start at the corner of the nose lock and move straight back to the tail, parallel with the center crease.

19B. Here's an alternate wing crease I like. Start at the corner of the lock and rake the fold down to the corner of the tail.

20. One wing fold is complete. Flip it over and make the other wing.

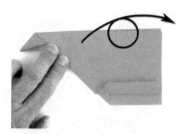

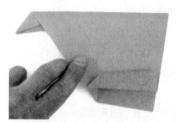

21A. Make the winglet about half the height of the fuselage.

21B. Flip the plane over and make the other winglet match.

21C. If you made the alternate wing crease, now make the alternate winglet to match the alternate wing fold.

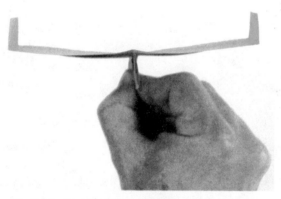

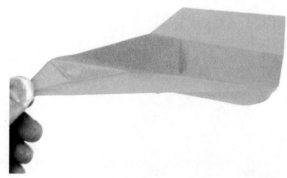

22. The finished Pro Glider. Note the slight positive dihedral. I love this nose lock. You'll see it on other planes in the book.

23. The striking look of the alternate wing fold. Note that the winglets are oddly shaped because they are still parallel with the main wing crease.

PRO GLIDER 2.0

This is one of a couple of variations on the wing-layering system with my new fuselage-locking technique. I've used the alternate wing crease from the original Pro Glider, but you can play around with making your own shapes and winglets.

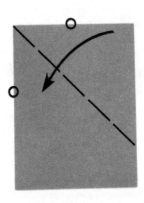

1. Start with the short side up. Make a diagonal fold by lining up the top edge with the left side of the page. Split the left upper corner in half with the crease.

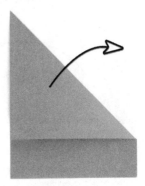

2. Unfold.

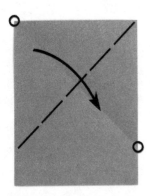

3. Make a diagonal fold in the other direction.

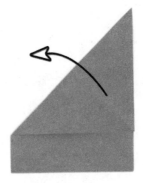

4. Unfold.

5. Flip the page over.

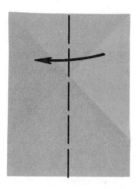

6. Fold in half.

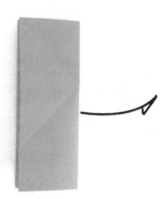

7. Unfold by pulling out the bottom layer.

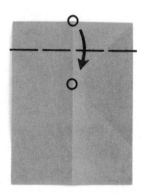

8. Fold the top edge down to meet the intersection of all the creases. Leave a little room to accommodate the creases to come.

9. Unfold.

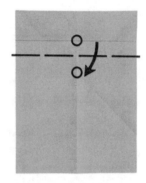

10. Line up the crease from step 8 with the intersection of the diagonals. This move spreads the CG back a little. You'll find this version may be better indoors. The Pro Glider may be better outdoors with a little up elevator. The magic is all right here.

11. Make a mountain fold along the crease from steps 8 and 9.

12. The layers should look like this. Recrease the diagonals now.

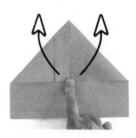

13. Unfold the diagonals.

14. Flip the plane over.

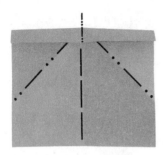 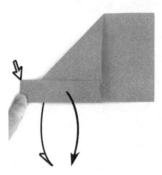 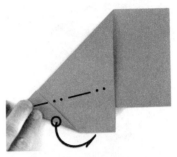

15. This looks complex. Check out the Pro Glider instructions for steps 12A through 12D on page 47 for this step if you're confused.

16. Pull the layers out and down while letting the corner move down. Have a look back at Pro Glider steps 13A through 13D on page 48 for more explicit directions on this.

17. Mountain fold this layer under. To do this, lift open the layer, then fold it in half by moving the edge marked by a circle down to meet the crease. (This is similar to steps 14 through 16 on page 48.)

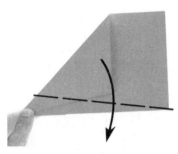  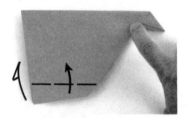

18. I'm going with the alternate version of the wing here, but use the original version (see step 19A of the Pro Glider on page 49) if you like.

19. Flip the plane over and make the other wing.

20. The winglet should be parallel with the main wing crease. The front of the winglet should be about half as tall as the front of the fuselage. Make both winglets while you're at it.

 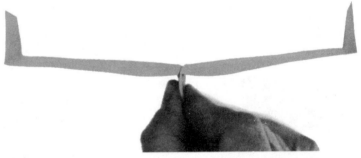

21. Both winglets are done. Open up the plane.

22. I really like the clean lines created by this folding technique. Keep the dihedral nearly flat and the winglets vertical.

BROAD DIAMOND

With the winglets folded, you get a diamond-shaped lifting surface. This is a sturdy, locked-together aircraft. Indoors, or even outdoors in some wind, you'll have a great time punching holes in the air with the Broad Diamond.

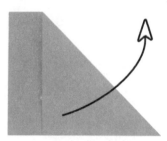

1. Start with the long side up. Move the left side against the bottom. The crease goes smartly through the corner.

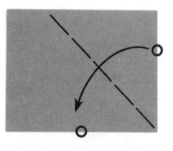

2. Unfold.

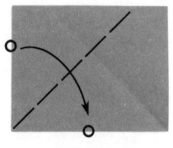

3. Now move the left side to the bottom.

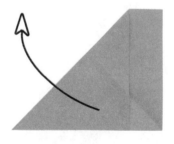

4. And unfold the left side.

5. Flip the page over.

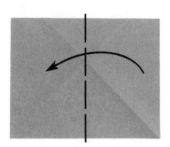

6. Fold the page in half.

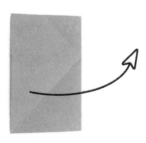

7. Unfold the crease from step 6.

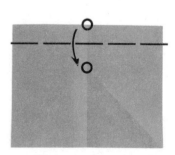

8. Fold the top edge down to the point where the creases meet. Stop just a little short of that point, no more than a millimeter.

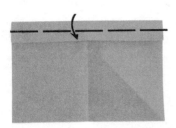

9. Fold the layered part in half.

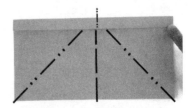
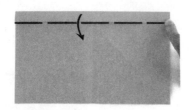

10. Fold the layered part down. This crease should hit the intersection of the creases perfectly if you left a little room in step 8.

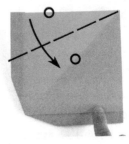

11. Here's the Collins nose-lock technique again. Collapse the plane together by making the mountain folds and valley fold shown. Refer to steps 12A through 12D of the Pro Glider on page 47 if you need help.

12. Pull the outer layers out and down. The marked point moves down to meet the wing layers. Again, refer to the Pro Glider (steps 13A through 13D on page 48) if you're confused.

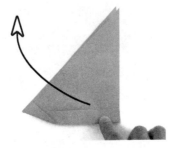

13. Lift the top wing to make the next couple of folds.

14. Line up the raw edge with the crease and make the valley fold.

15. Fold the front edge over to meet the crease.

16. Remake the long crease.

17. Flip the plane over and repeat steps 13–16 for the other side.

18. Make the wing crease start at the edge of the locking layer and slope down to the lower corner of the tail.

19. Flip the plane over and make the other wing.

20. Make a winglet by folding up just short of one-third of the wing, parallel to the main wing crease.

21. Flip it over and make the other winglet match.

23. Your freshly fabricated Broad Diamond. Keep a low dihedral angle and give it an easy shove.

22. The folding is finished. Open up the wings and winglets.

MAX LOCK

This is the maximum size, true delta wing I could muster with the new nose-locking idea. It's certainly maximum cool. Look at that nose just blending with the fuselage. It's nearly seamless. Gotta love that!

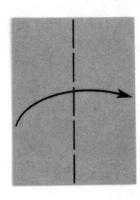

I.   Start with the short side up. Fold in half.

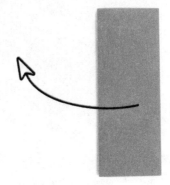

2.   Unfold.

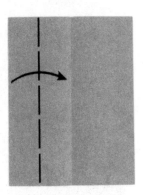

3.   Fold the raw edge to the center.

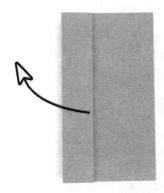

4.   Unfold.

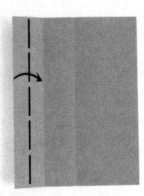

5.   Fold the raw edge to the crease from step 3.

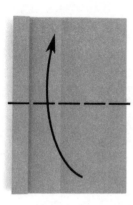

6. Fold in half.

7. Unfold.

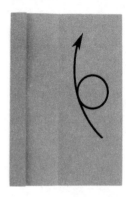

8. Flip the plane over.

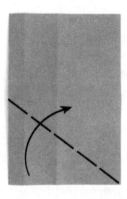

9. Fold a diagonal that starts at the center crease and hits the lower right corner.

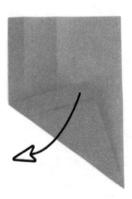

10. Unfold.

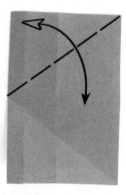

11. Fold and unfold the other diagonal.

12. Flip the plane over.

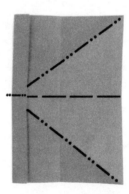

13. The Collins nose lock is now commencing. If this is your first attempt at the technique, you may want to review the Pro Glider instructions in steps 12A through 12D (page 47) and 13A through 13D (page 48).

14. Locking sequence continues with pulling the layers down and out and forcing the corner to line up with the layers below.

15. Open the top layer to complete the next couple of folds.

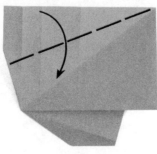

16. Fold the raw edge to meet the long crease.

17. Fold the front edge over to meet the crease.

18. Remake the long crease.

19. Flip the plane over and repeat the sequence of steps 15–17 for the other side.

20. Make a wing crease that starts at the corner of the locking layer and rakes down to meet the rear corner.

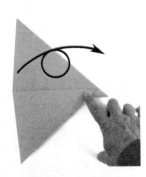

21. Check out that wingspan! Flip the plane over and make the other wing.

22. Keep the winglets to under one-third of the wing to preserve all that wing width.

23. Another very clean, elegant design with a slight dihedral and amazing performance.

LF-WHAT?

This is a hybrid version of the LF-1 from my book *Fantastic Flight*. I mixed in Toda's nose-locking technique from his world record duration plane. This is a great glider and really fun to fly.

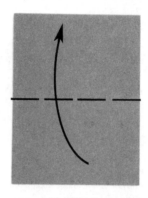

1. Start with the short side up. Fold the bottom up to meet the top.

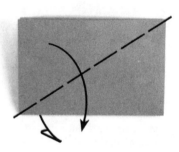

2. Make a diagonal fold across the top layer. Make a matching one on the back layer.

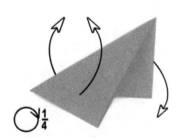

3. Unfold the creases from step 2. Unfold the center crease by pulling out the bottom layer. Rotate a quarter turn so that the long side is now on top. Wow, I could've just said unfold everything and make sure the center crease is a vertical mountain fold.

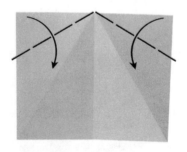

4. Fold the top edges down to meet the creases. Leave about a crease width of space to make the layers fit together later.

5. Fold the raw edges over to meet the crease. Again, just a little breathing room there is good.

6. Prepare to use up the breathing room. Remake the long creases.

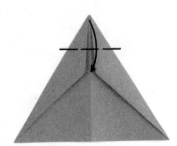

7. Fold the top down to meet the two corners.

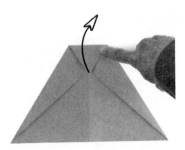

8. Okay, the nose doesn't actually touch the two corners. You're right. Just unfold it now.

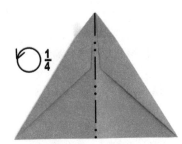

9. Mountain fold in half and rotate a quarter turn to the left.

10A. I'm stealing from two of the best for this plane. The basic design here is the LF-I from my book *Fantastic Flight*. The nose lock we're about to start is Takuo Toda's (as far as I know). He's the guy who holds the record for paper airplane flight duration.

10B. Squash this point, using the vertical crease as the limit.

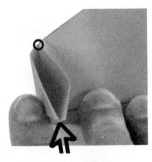

11A. Alright, here's hint for a squash fold. Lift up the layers. The little circle is marking the end of the vertical crease and the place where you're about to squash the nose. Open up the layers and push the nose down.

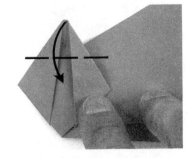

11B. When you flatten it all out, you get the photo above. Fold the top point down to the center.

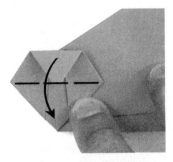

12. Fold across corner to corner, moving the top half all the way down.

13. Move the left half of the squashed part behind and there you have it: Toda's locking system.

14A. This photo is the completed lock.

14B. Here it is zoomed out to show you the wing crease. Start at the corner of the lock and rake down to the rear corner.

15. Flip the plane over and make the other wing.

16. This photo shows both wings made.

17A. In this photo, I let the top layers spring open to make a winglet that will point down. Fold up about one-third of the wing.

17B. Flip the plane over to make the matching winglet.

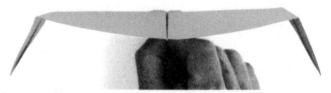

18. Here's the completed LF-What. Check out the dihedral and the winglets slightly leaned outward. Spiffy nose lock, huh? Once I saw this lock, I invented my own for the Pro Glider. Now you've got options!

LOCKED AND LOADED

This is a very fast, fun plane. I like the wing shape and locked fuselage. You can get away with 24lb paper on this one, particularly if you're going for serious distance with style.

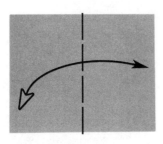

1. Start with the long side up. Fold in half and unfold.

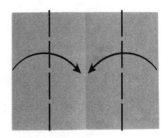

2. Fold the outside edges to the center.

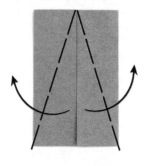

3. Fold the raw corners back out, on the diagonals.

4. Fold these corners down, following the raw edges.

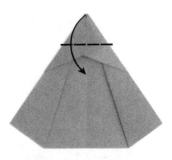

5. Fold the nose down. It lines up with the where the top layers end.

6. Unfold the nose.

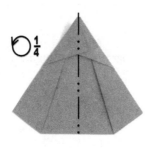

7. Mountain fold in half and rotate a quarter turn left.

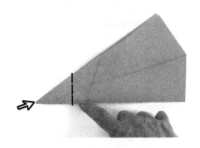

8. Squash fold the nose.

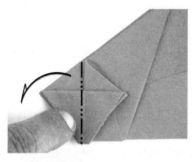

9. Here's a closer view for you. Fold the top down across the two corners of the squash.

10A. Fold the left half of the squashed assembly behind with a mountain fold.

10B. This photo is the completed locking sequence.

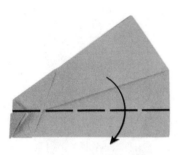

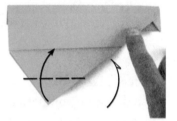

11. This photo is a zoomed-out view to better show the wing fold. Start at the corner of the lock and proceed straight back to the tail parallel with the center crease.

12. Flip the plane over and make the other wing.

13. Make the winglets by moving the corner of the wing up to meet creased layer. Do both sides.

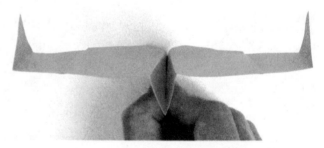

14. Both winglets are done. Open up the wings.

15. The finished Locked and Loaded plane. A fairly flat dihedral will work with these locked fuselage designs because they don't flop open in flight. I like the lines on this one. The heavier wing loading accounts for the last part of the name.

ULTRA GLIDE

There's just something about the way this one flies. The very flat wing layers, locked nose, and clean over-all design create a nearly perfect plane. I'd take this into duration competition without hesitation. With the right paper and the right throw, I'd like to see what this one would do.

1. Start with the short side up. Fold the paper in half both ways and then flip it over so you've got mountain folds.

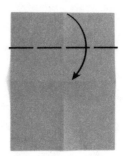

2. Fold the top edge down to the center.

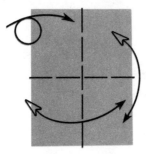

3. Again, fold the top edge to the center and then unfold.

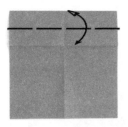

4. Fold the top edge down to the crease from step 3.

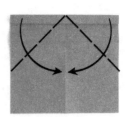

5. Fold both corners down to meet the center.

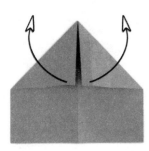

6. Unfold.

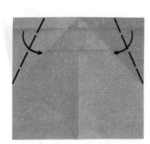

7. Fold the raw edges to the creases from step 5.

8. Remake the creases from step 5. Tuck the marked corners under the marked raw edges as you make the creases.

9. Fold the top point down to the corners.

10. Unfold the top point.

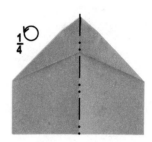

11. Mountain fold the plane in half and rotate one quarter turn to the left.

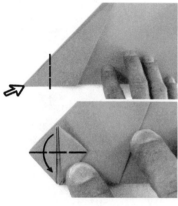

12. Here I'm leaning on the Toda's locking system again. Squash fold the nose and then bring the top point down to the bottom.

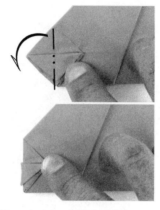

13. Move the left side of the squash assembly behind. The bottom photo shows the finished move.

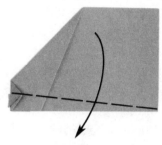

14. The wing fold starts at the corner of the lock and rakes downward to the rear corner.

15. Flip the plane over and do the other wing.

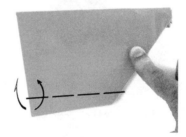

16. Make the winglets as shown.

17. There's a bit of similarity in this line of planes that all use the new Collins nose lock. You'll find the one that pleases you best and stick with it. There's enough subtle weight shifting in the designs to match your folding style.

SUZY LOCK

The weighting system is borrowed from Suzanne, the world-record plane. The nose-locking system is taken from Takuo Toda's Sky King. The wings are just basic glider design. This is a really good paper airplane for all its found-technology beginnings. If you're not in the mood for a long folding and taping session required for the record breaker, try this much more accessible (at least in the United States) design that uses similar techniques. Suzy Lock easily defeats most competitors for distance or duration in the amateur ranks.

1. Start with the short side up. Fold and unfold the upper diagonals through the corners.

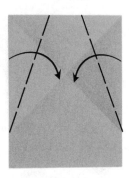

2. Fold the raw edges in to line up with the diagonal folds.

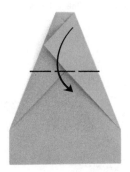

3. Fold the top down across where the diagonals meet. Make sure the raw edges align with the creases from step 1.

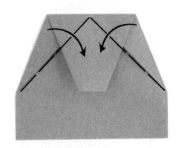

4. Follow the existing creases.

5. Fold over one-third.

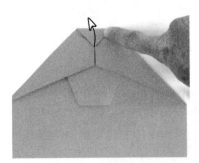

6. One-third means the height of the whole layered part will be about the same size as the flap part below it after the fold is made. Now, unfold.

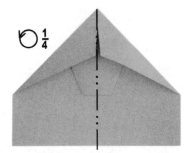

7. Mountain fold the plane in half. Rotate one quarter turn to the left.

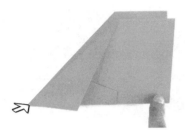

8A. Squash fold the nose (see page 30) . . .

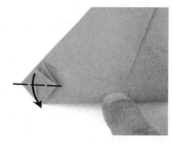

8B. . . . and then fold the top layer down.

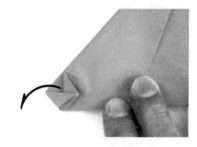

9A. This is a close-up view to show how the left side of the squash gets folded behind.

9B. Here's the result.

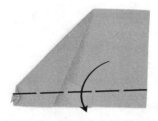

10. The wing fold starts at the corner of the lock and rakes upward slightly as it goes toward the tail.

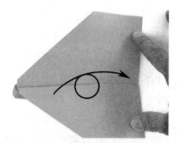

11A. We'll flip it over and fold the other wing, but first take a peek at how tall the fuselage turned out to be.

11B. Okay, flip it and keep working.

12A. In this photo, both wings are folded. Let the top layer spring open to make the downward turned winglet.

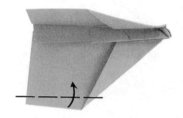

12B. The front of the winglet is about half the width of the fuselage.

13. Flip the plane over and make the other winglet match.

14. This is a pretty good glider, using the basic Suzanne layering for weight and the Toda nose lock. You'll need some up elevator to keep it flying flat. You could play with making the fuselage taller, and getting rid of the winglets. I like the look of this version, though, and it's a really good glider.

BASS ACKWARDS

The Bass Ackwards is an odd-looking but great flying design. The locked fuselage and tall winglets will give you fun flights. The "backwards" looking form is designed to raise questions anywhere you fly.

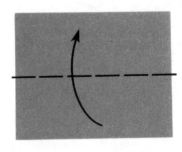

1. Start with the long side on top. Fold in half.

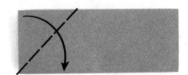

2. Fold one corner down.

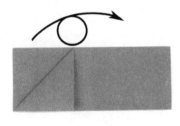

3. Flip the plane over and fold the matching corner down.

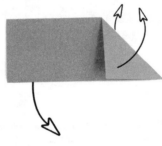

4. Unfold the whole plane. Leave a mountain fold down the center.

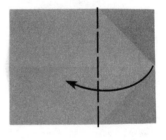

5. Make a crease that connects the ends of the two creases from steps 2 and 3.

6. Unfold the crease from step 5.

7. Fold the right edge over to meet the crease from step 5.

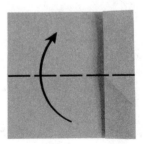

8. Fold in half, going over the existing crease.

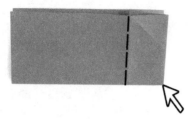

9. Squash this corner. (Refer to page 30 for instructions on how to make a squash fold.) Use the crease marked with the valley fold to help.

10A. Fold the two corners down. Make sure the raw edges meet the center.

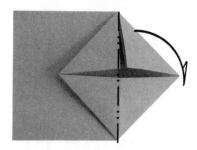

10B. This picture shows the result close-up. Fold the right side behind.

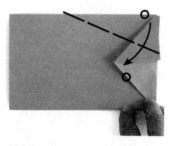

11. Make a valley fold so that the top marked edge lines up with the marked edge of the pocket.

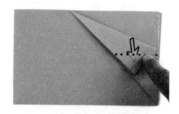

12. You guessed it. The corner goes into the (currently covered) pocket. The dotted line reveals the top edge of the pocket.

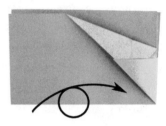

13. Flip the plane over and repeat steps 11 and 12 for that side.

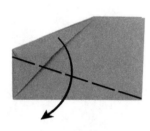

14. The main wing crease goes from the front corner of the pocket to the rear corner of the plane.

15. Flip the plane over and make the same wing crease on the other side.

16. Make a nice, fat, winglet crease (nearly one-third of the wing) parallel to the wing crease.

17. One down, one to go. Make the other winglet match.

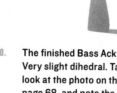

18. The finished Bass Ackwards. Very slight dihedral. Take a look at the photo on the top of page 68, and note the shape of the body in profile. It's exactly backward of the classic dart.

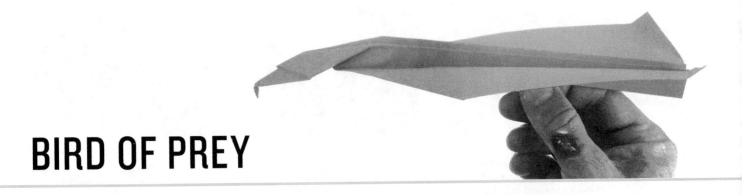

BIRD OF PREY

The striking profile is only the beginning. A true cambered wing is nearly impossible to fold—until now. The Bird of Prey is where totally awesome design meets technical achievement. This was an accidental leap forward in paper airplane design. I was only going for the eagle head shape. The particular bunching of paper, combined with my choice of wing folds, results in a reliably cambered shape. After decades of trying to do that on purpose, it happened when I was playing with the origami instead of the aerodynamics.

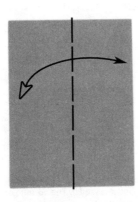

1. Start with the short side on top. Fold the page in half and then unfold.

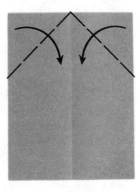

2. Fold the top corners to the center.

3. Fold the top point down to meet the two corners.

4. Now we're going to squash the marked corner. You may find it easier to first make a valley fold that will act as a guide for the whole step. To do this, bring the marked corner down to the center crease and press down. (I like to just go for it without that preliminary step.) To finish the squash, unfold the valley, make a mountain fold along the indicated line, then press everything down to secure.

5. There it is. Squashed! Move the corner back to the right. Just follow the existing crease.

6. Once more, with feeling! Squash the corner, using the same process as in step 4. (Start with that valley fold if you need to.)

7. And move the little corner back to the left. The bottom photo shows the whole move, finished.

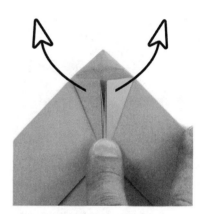

8. Time for a petal fold. (If you need help, refer back to page 31.) The bottom shot is a close-up. To get ready to petal fold, fold the edges to the center.

9. Fold the top triangle down, then unfold.

10. Unfold the two flaps.

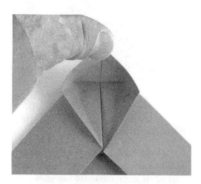

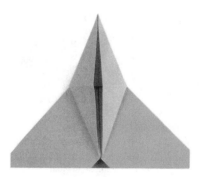

11. Enough warm-up. Let's petal fold. Lift the top layer, using the crease from step 9 as the limit.

12A. This photo shows the layer getting lifted. Keep lifting, and follow the creases on the bottom. To complete the petal, reverse the direction of the creases on the top so the whole thing lies flat.

12B. This photo shows the completed petal fold.

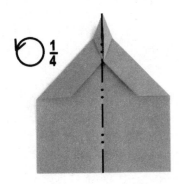

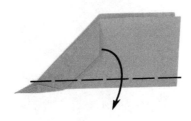

13. Fold the whole plane in half. It's a mountain fold, which means you flip the plane over and make a valley and then flip it back over. The next picture shows the plane rotated a quarter turn to the left.

14. Make the wing crease so that it just touches the top corner of the petal fold. Make the crease parallel with the center crease.

15. Make the other wing crease.

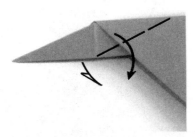

16. Now make an outside reverse fold. (Refer to page 29 for help.) Notice that the crease is parallel with the top of the nose. Also notice that the end of the fold meets the end of the vertical crease. Just spread the nose apart and pull the edges down around the nose.

17A. Wow! How cool is that? You just made a raptor's head. This photo indicates a small inside reverse fold to create the illusion of a curved beak.

17B. This photo shows the result.

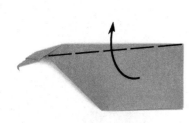

18. The main wing crease gets folded upward on this design. Start at the front where the head meets the front of the wing and end at the upper rear corner.

19. One wing done. Notice the layers bunching a little. Let them. All that petal and reverse folding stuff is working in a weird way for you. Flip it over and make a matching wing fold.

20A. This shows the wing folds completed.

20B. Let the plane flop open, with the top of the wings showing. Make a couple of big elevator folds. This is an extraordinary step for a paper airplane. Usually that much up elevator means I really messed up the design.

21. Still more weirdness: Hold the head firmly with one hand. Grab the tail feathers where all the layers come together. Now pull, gentle but firm. That helps set the camber.

22. Check the curve of the wings in this photo. Look at the big gap in layers under the wing. That's good for this design. The photo at the top of page 70 shows this bird in profile. What a cool design! Work with the elevator controls to get a smooth glide. It rides the air like few folded planes can.

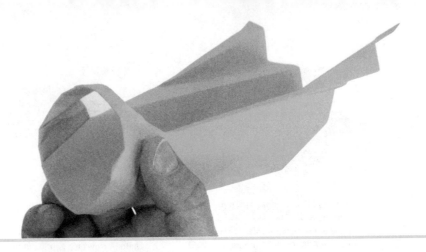

RING THING

You knew there was going to be at least one plane in the book that made you beat your head against the wall when you tried to figure it out. This is the plane. You also knew you'd have to try. The crazy thing is that it actually flies pretty well.

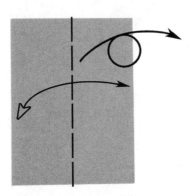

1. Start with the short side up. Fold the page in half, unfold, and then flip the page over. This is alternatively known as mountain folding the page in half. See how step 2 starts with a mountain-like ridge running down the center?

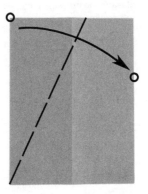

2. Folding through the bottom left corner, move the left upper corner to the raw edge so that the crease will hit the lower left corner.

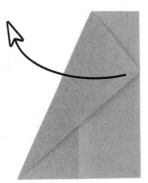

3. Unfold.

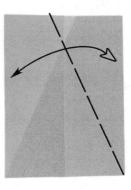

4. Repeat steps 2 and 3 on the right side.

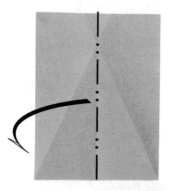

5. Follow the center crease mountain fold to fold the left side behind.

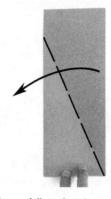

6. Danger follows from here on out. We're making a waterbomb-like base with the next few folds. Start by following the existing crease here and just letting the top spring open.

7. Move the top layer back to the right, lining up the valley fold with the center crease. Start your valley fold at the marked point, where all the other creases meet.

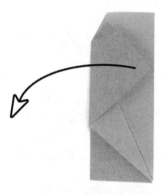

8. Great work! Now unfold the whole thing. Make sure your paper is facing the right way: the center fold should still be a mountain.

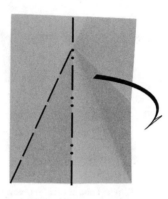

9. Move the right side behind by following the center mountain fold and then following the marked valley.

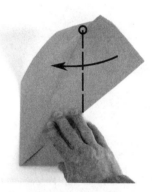

10. Move the top layer back to the left, making a valley fold that follows the center crease as in step 7.

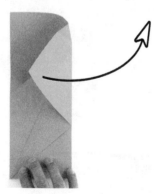

11. Unfold the whole thing. I know, fold and unfold, make up your mind, right?

12A. Let the page warp upward, which it should do if you haven't smoothed out your creases. Press down on the point where all the creases meet.

12B. I'm pointing to it and starting to press down.

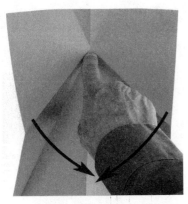

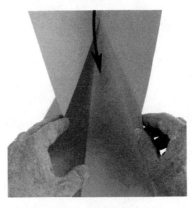

13A. It "popped" the other direction! This only works if your creases all meet in that one spot. Careful folding is the key. If yours doesn't pop, it'll be a little harder to assemble it like these photos, but you can do it.

13B. Start by bringing the sides together, and then bring the top down.

13C. This photo shows me holding the layers flat and then pressing the very top of the plane down. You want the top part to flatten out evenly. The resulting creases need to go through the left and right corners, marked here with circles.

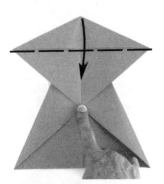

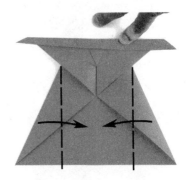

14. Admire your work for a moment before folding the top triangle down along its raw edge base.

15. Fold just one-third of the way down the top triangle.

16. Fold the corners inward. The creases are parallel with the center crease and start, on the upper end, where the raw edge meets the crease edge.

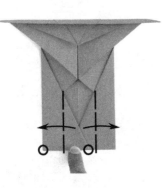

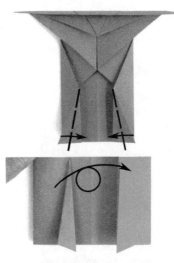

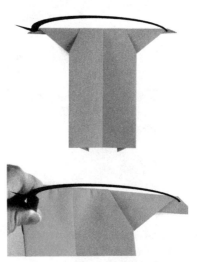

17. Fold the corners back out. The bottom raw edges are folded in half and the raw corner should meet the creased edge as marked.

18. Fold the little triangles in half, but this time line up the crease edges. The bottom photo shows a close up of the completed fold. Take a deep breath, flip the plane over, and get ready to fold a circle.

19. Tuck the right side into the left side. This is made easier by starting to curve the top edge. Just bend it into a circle, flexing that layered part until it stays curved a bit. Open up the left arm and stuff the right arm in there. It should slide in without too much struggle.

20A. Now for the tough part. Hang on to the overlapping part while you pinch all those layers together with a mountain fold. The next photo shows the work in progress. Keep folding and pinching it together, and then rock the layers forward by folding the front of the plane to the inside of the circle just a little bit.

20B. Don't go crazy. This step generates the most cries of angst and general cursing of the Paper Airplane Guy. The arms will pop open before you finish. It's hard to get the ring layer flat. It's difficult to push in only a little paper in the front of the plane. I know that. I'm warning you now.

20C. Hands free on the completed ring in this photo. See, it's possible.

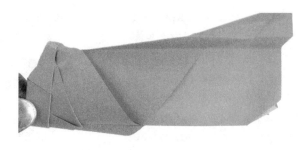

21. This photo shows your next tasks. Reverse fold (page 29) the rear lower corner about half way up the height of the fuselage. Then bend in a little up elevator in the upper rear corners of the wings.

22A. The finished Ring Thing.

22B. Amazing! But wait, there's more . . .

23. Bonus points for turning the Ring Thing into "Love at First Flight." To fly this plane, wrap your hand around the plane and hold it loosely, a little like holding an American football. Give it a shove forward. Try a level toss to start.

THE BOOMERANG

The Boomerang has become an international favorite. When thrown correctly, the plane will make a full circle and return to you! Unexpectedly for me, this plane will flap its wings quite well when folded with A4 paper. The A4 paper changes the geometry of the layers just enough to create an articulating leading edge instead of a solidly thick one (like US 8½ by 11, letter size creates). By removing 19 mm from the long side of US letter size, a sheet with the A4 ratio will be created. I was pleased and surprised to see the plane used both ways, flapping and circling, in the 2012 Red Bull Paper Wings contest in Austria.

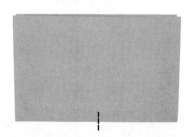

1. Start with the short side up. Fold the paper in half.

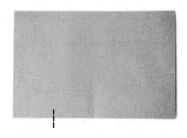

2A. Make a pinch mark at midpoint of the crease.

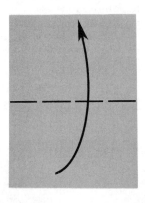

2B. Here's how to make the pinch. Move the right lower corner over to the left lower corner. Use one finger to press firmly on the creased edge. Then unfold.

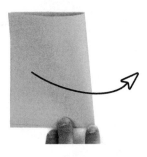

3. Now make a pinch at the halfway point between the left lower corner and the first pinch.

4. Unfold.

5. Make a crease that goes from the pinch from step 4 up to the upper left corner.

6A. Squash this corner. (Page 30 has instructions for the squash fold.) Start by opening the long raw layers.

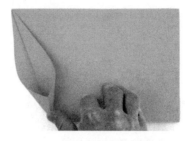

6B. Now that the layers are open, press the creased edge down, lining it up with where the layers below meet.

7. The completed squash. Following the existing crease, fold the flap behind.

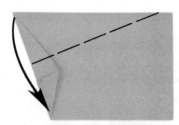

8. The upper left corner will eventually get tucked into the pocket formed by the squash fold. To start the process, fold the upper left corner down. Stop short of the lower left corner. Make sure the creased edges line up.

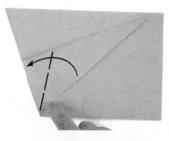

9A. This photo shows the crease needed to narrow the corner enough to go into the pocket. The crease should follow the pocket shape and stay just inside the creased edge to ensure it will fit.

9B. This photo shows the flap just a little short of the bottom of the pocket.

10A. Here's how that completed crease should look. Now it's time to put the corner into the pocket.

10B. A wider view shows how the layer on top is warping. We'll fix that in a moment. First, get that corner into the pocket.

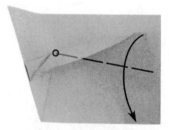

11. The corner is in. Now let's deal with the warping issue. The crease will start at the end of the narrowing crease, marked, and will let the upper right corner swing down close to the center crease. Nail down the fold.

12. One side down; one to go. Your corner may be a bit more off the center crease than mine. That's okay. Just make the other side match. Make steps 8 through 11 on the other side.

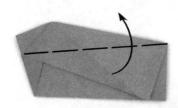

13. Lift the top layer, making a crease that starts at the edge of the pocket and extends to the upper right corner.

14. Flip the plane over and make the same crease on the other side.

15. The whole front assembly is about to get squashed flat. Start pushing here.

16. Let the whole plane open up along the center crease. Keep pushing the front toward the center crease.

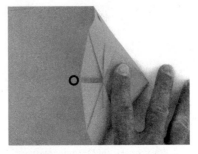

17. The center crease is flat now. Keep coming down. Before nailing the squash down, adjust the marked gap so that corners stay close to the center crease.

18. Nice and flat. Flip the plane over.

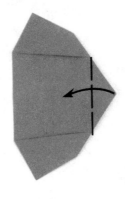

19. Make a crease that connects points where the layers meet. The point should fold over to meet the center crease.

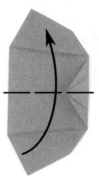

20. Remake the center crease.

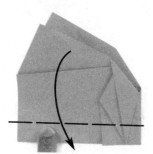

21. Make a wing crease parallel to the center crease.

22. Here's a close-up of that wing crease. Note the small triangle at the nose.

23. Flip it over.

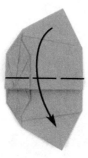

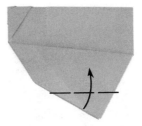

24. Make the other wing crease. Get those wing tips lined up.

25. Make a winglet parallel with the wing crease.

26. You can see the corner is a little short of the end of the layer. I've made winglets that touch the layer. That works, but the plane flies fine with shorter winglets. Make the other winglet match.

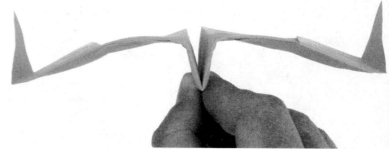

27. Bend some curve into the leading edge, close to the fuselage.

28. Above, the finished Boomerang. Note the wing droop. That's negative dihedral angle. This allows the plane to stay in a banked turn when you launch it leaned over on its side.

29. At the right, the proper launch angle. Positive dihedral angle creates an aircraft that rocks back to neutral when it launched at this sideways angle. With positive dihedral, the center of gravity is further below the center of lift, creating a sort of pendulum effect. Negative dihedral defeats that ability of the plane to right itself. Extra up elevator will make the circle it flies smaller.

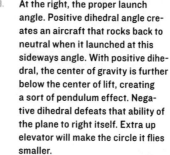

THE STARFIGHTER

The Starfighter is a two-time international distance champ, topping out just short of 100 feet. That's pretty wimpy by world-record or even Red Bull Paper Wings champions standards. The biggest trick this plane managed was surviving international shipping in good flying condition. The hexagonal wing shape was the winning idea.

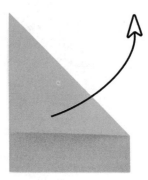

1. Start with the short side up. Make a diagonal fold. We're starting a waterbomb base (page 27), so if you know how to do that, skip ahead to step 10.

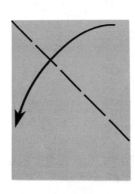

2. Unfold.

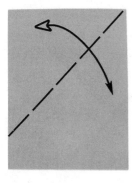

3. Make a diagonal fold in the other direction.

4. Unfold.

5. Flip the paper over.

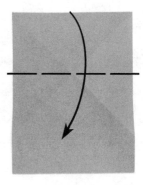

6. Fold the big X in half by bringing the crease corners down to where the diagonals meet the raw edges.

7. Partially unfold this crease and stand the paper up with this crease making the top of a lopsided tent.

8. Push down where all the creases meet until the paper "pops."

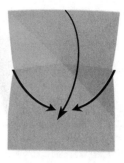

9. Bring the sides in and the top down by following the existing creases.

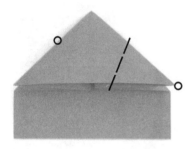

10. The completed waterbomb base. Move the marked corner over to meet the creased edge. Do this in a way that allows the creased top edge of the flap to end up parallel with all the horizontal raw edges in this photograph.

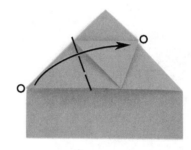

11. Note that creased edge of the flap creates a perfect half of a square above it. Now bring the left corner over to meet the marked corner on the right.

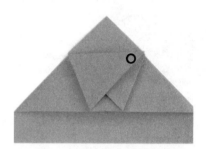

12. Lift up the marked flap and spread apart the raw edges to create a pocket.

13. Now put the other flap inside the pocket.

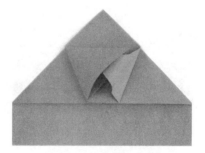

14. Here's the flap going in. Put it all the way inside so that both flaps will lie flat.

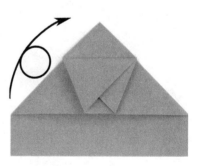

15. Flip the plane over.

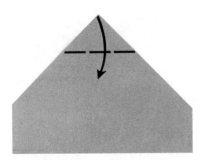

16. Fold the top down using the corners of the flaps on the other side as the limit.

17. Flip the plane back over.

18A. We're folding the plane in half in a really weird way. Allow the joined flaps to lift up as the other surface is folded.

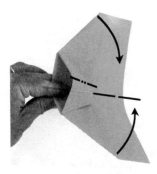

18B. Another angle showing the folds starting to happen. Take care to keep the flaps locked together.

18C. Looking good. Now just make those creases. Make sure the raw corners line up as you make the long crease down the center.

19. Nice. Let the layers open a little.

20A. We're about to make two new creases on the tunnel. These valley folds start at the rear corner of the tunnel and extend forward, parallel to the center crease. The whole crease that is the top of the tunnel moves down.

20B. I pinch the top of the tunnel together to help with this move.

20C. Here are the finished valley folds, which run parallel to the center crease.

21. Nail those creases and flatten the whole thing.

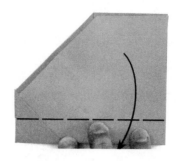

22. The main wing crease is parallel with the center crease. Use the creases from the last step as the guide for wing-crease placement.

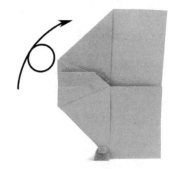

23. One wing-crease finished. Notice how the crease is lined up with the tunnel crease. Flip the plane over and make the other wing crease.

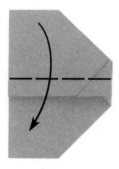

24. Make the wings match.

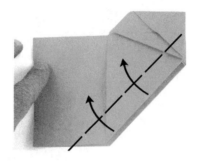

25. Fold the leading edge over. The valley fold should be parallel to the leading edge crease. We're getting ready to reverse fold a corner of the tunnel.

26. Notice how the corner of the tunnel is lined up with the wing crease where marked. Now unfold.

27. Here's the corner we'll reverse. If you've never done a reverse fold, the next couple of steps will guide you through this version. Start by opening up this corner and following the long crease from step 25.

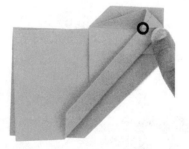

28. Make sure the center of the corner is lined up with the wing crease where marked. Take your time here.

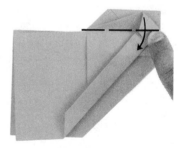

29. Now bring the whole flap back down.

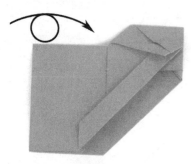

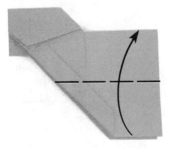

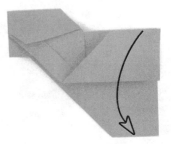

30. Great. You've done an inside reverse fold! No time to rest. Flip the plane over and do the other wing by repeating steps 25–29.

31. Fold the raw edge up to meet wing crease. We're starting to form the hexagonal-shaped wings.

32. Unfold.

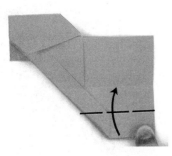

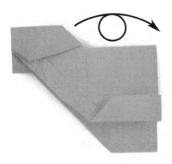

33. Fold the raw edge to the crease from step 31.

34. Flip the plane over and make the same creases from steps 31–33 on the other wing.

35. Open up the wings and adjust them so they look like the finished photo.

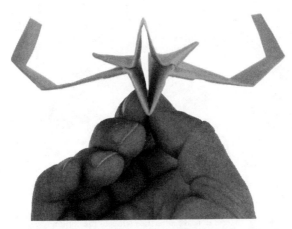

36. The finished Starfighter. Note the slight negative dihedral on the main wing crease. That's part of the striking appearance. The tunnel through the nose is a very nice feature too.

FOLLOW FOILS

The next four paper airplane designs are for **FOL-LOW FOILS**—a crazy kind of paper airplane that will stay aloft for far longer than the current world record for duration. The trick is to hold a piece of cardboard behind the plane to generate a steady stream of upward moving air. The airplane rides in this air as you walk holding the cardboard. As long as the air is moving upwards at the same speed as the plane is gliding downward, you can keep the aircraft from losing altitude. It's sort of like sky surfing. You walk behind the plane, creating a wave of air, and fly it around a room, a gym, or airplane hangar.

In fact, I flew one of my follow foil models for more than 30 minutes at the Exploratorium in San Francisco. The world record for duration was under 28 seconds at the time. Shortly after I submitted this record-breaking attempt to Guinness for consideration, they changed the rules to forbid Follow Foil–type planes. Later I found out that two Brits had tried the same thing that year. In retrospect, I get Guinness's point. Allowing Follow Foil designs into the record book would fundamentally change the duration contest from an aerodynamic challenge to a human endurance challenge.

ADJUSTING AND LAUNCHING A FOLLOW FOIL

First, get the plane gliding in a straight path by making control surface adjustments. If the plane turns right, add some drag to the left side of the plane by increasing the up elevator or leading edge droop. You could also take away some elevator or leading edge droop on the right to shave drag and help speed up the right wing. Adding drag to the faster wing or subtracting drag from the slower wing will eventually give you the straight-line flight you need. Use your elevator adjustment skills to get a flat glide path. Eliminate a tendency to climb or dive. A smooth, straight flight is what you need before you practice launching.

To launch a Follow Foil, I spread the fingers of my left hand, palm up. I position my left hand about shoulder height. The plane rests near my fingertips so that air can pass through my open fingers and lift the plane. I hold a sheet of cardboard in my right hand, angling it so that the top leans back toward me. This way, the angled cardboard scoops up air when I start walking forward. I walk forward with the plane in my left hand and the cardboard in my right, adjusting my speed and the cardboard position until I can make the plane lift off from my fingertips. Once the plane is up, I hold the cardboard with both hands to help control the flight.

Turning is simply a matter of angling the cardboard to push more air under one side of the aircraft. Pushing more air under the left wing will cause the left side of the plane to lift more, and turn the plane to the right. It's a little like causing an aileron roll. To turn right, force more air under the left wing. To turn left, force more air under the right wing.

Let the air come up through your fingers to lift the airplane off your hand.

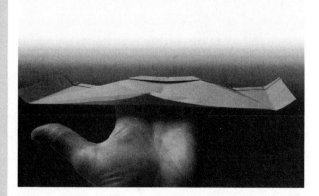

A carefully angled piece of cardboard generates the airflow needed to keep the plane aloft.

LF-FF

The Locked Fuselage–Follow Foil, or LF–FF, is perhaps the best dual-purpose design, bridging the paper airplane world and the follow-foil world seamlessly. With standard paper, you'll get slow motion–like glides across the room. When made from phonebook paper, you'll get a very serviceable follow foil. I really like the locked together fuselage and super light wing loading.

This is among the fastest of the follow foils, so if you're looking for that perfect plane to fit into your aerobics routine, this one is for you.

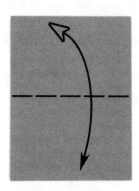

1. Start with the short side up. Fold the page in half and unfold.

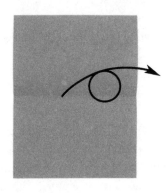

2. Flip the page over.

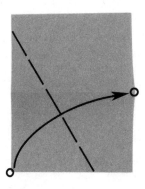

3. Fold the lower left corner to the marked end of the center crease.

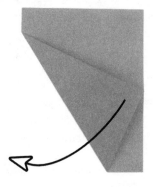

4. Unfold step 3.

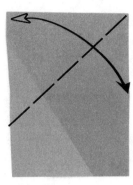

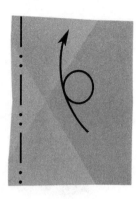

5. Fold and unfold the upper left corner like step 3.

6. Mountain fold one-third of the distance between the left edge and the intersection of all the creases. Flip the page over.

7. If you hit a perfect one-third in step 6, the creased edge will hit the intersection. It's worth working on that crease. It's okay to land short of one-third. If you land long, the layers will scrunch up as you put the creased edge on the intersection.

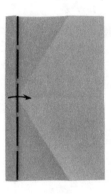

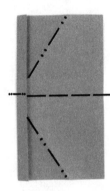

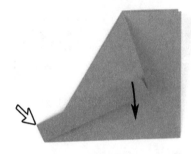

8. Make a valley fold that goes right through the intersection of the creases.

9. This is the Collins nose-locking system, as explained in the Pro Glider steps 12A through 13D on pages 47–48. Note that unlike the Pro Glider, the layered part will *not* align with the center crease.

10. Continuing with the nose-locking sequence, pull the outer layers down and out, until the marked corner gets pulled against the two creased edges that will become the wings.

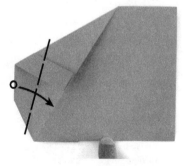

11. The last step was a little weird. Check out the Pro Glider instructions in steps 13A through 13D (page 48) for a more explicit set of steps to make that happen. Lift up this side.

12. Fold the raw edge to the crease.

13. Fold the marked edge to the crease.

14. Now remake the crease.

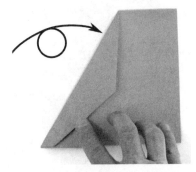

15. Flip the plane over and do steps 11–14 on the other side.

16. The wing fold starts at the corner of the locking layer and rakes down to the corner of the tail.

17. Flip the plane over and make the other wing crease.

18. The winglet crease is parallel to the main wing fold. The trailing edge is about the same height as the nose. It's okay to be a little taller than the nose, but don't go shorter.

19. One winglet is finished. Fold the other one.

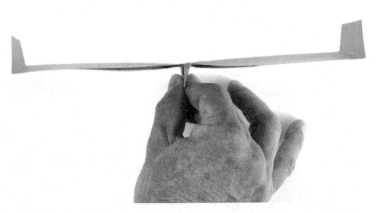

20. The finished Locked Fuselage–Follow Foil. Super light wing loading allows this plane to glide very slowly when made from regular paper. The real fun is the phonebook version. Give that a try.

FFF-1

I love this plane. The Folded Follow Foil-1 looks all awkward angles and rough edges. It reminds me a bit of the Air Force's F-117A stealth fighter in that regard. To adjust it, you can play with the winglets or big diagonal folds to add more drag to the side opposite the turning direction. The glide ratio (see page 12) is amazingly good on this plane. It's a nearly perfect follow foil when made from phonebook paper. This is one of the first planes I invented for this book. I've had to wait years to put it in the paper airplane shows. I found that people are irate if you show a plane for which they can't buy the instructions.

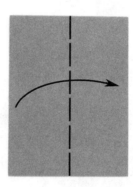

1. Start with the short side up. Fold the page in half.

2. Fold the top layer in half—but leave the bottom layer as-is.

3. This is a bit tricky. Notice that the valley fold dashed line is faded out a little. That's because the fold is happening on the layer below. Pull the top layer to the left. The marked creased edge on the right (which is on the top layer) will line up with the marked creased edge on the left (which is on the bottom layer). To do that, the layer under the top layer gets folded in half.

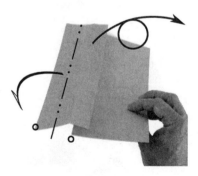

4. See how step 3 turned out? Now mountain fold the leftmost section in half. Note that the marked corners will end up touching. Flip the airplane over after this fold.

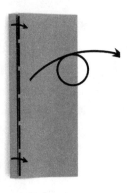

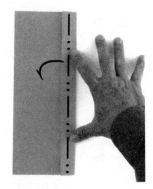

5. Fold the layered part in half.

6. Fold the layered part to the right. Flip the plane over.

7. Mountain fold the layered part in half, tucking the left edge underneath.

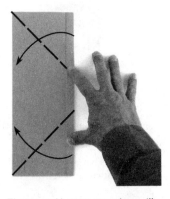

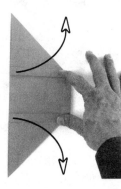

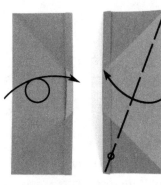

8. The top and bottom raw edges will line up with the raw edge on the left. The corners get split on the diagonal.

9. Unfold step 8.

10. Flip the plane over.

11. The raw edge on the right lines up with the diagonal crease. Don't crease through the layers. The marked point is as far as you should go. It's not necessary to crease all those layers.

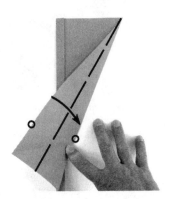

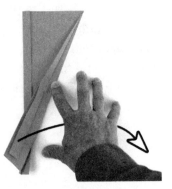

12. Fold the raw edge back to the crease from step 11.

13. Fold the raw edge back to the crease from step 12.

14. Unfold steps 11 through 13.

15. Now for the other corner. Repeat steps 11–14 for the lower corner. The middle fold is a mountain when viewed from this side, but you're just folding back and forth with valley folds.

16. Make a couple of winglet folds that are about one-third wider than the layered part.

17A. Unfold step 16.
17B. The right photo shows the result.

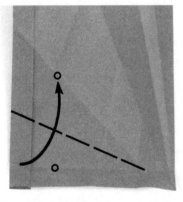
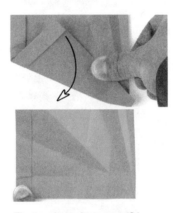

18. Line up the winglet crease with the diagonal crease from step 8.

19. The top photo shows step 18 happening. Notice that the lower right corner is not flattened. Just let it curl. Then unfold step 18. The bottom photo shows the result. Do the same thing to the top wing.

20. The left photo shows both sides done. The right shows the next move starting. Accordion the folds from steps 11–15 by pushing the raw edge to the left.

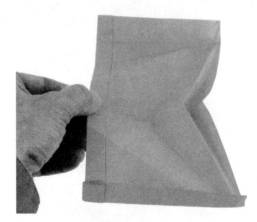

21. The top photo shows the accordion folds coming together nicely. The bottom photo shows how to pinch the center together to put a center crease in the accordion without creasing the leading edge.

22. The finished FFF-1. This is one of the real gems in the book. This is a very good follow foil.

ROGALLO FOLLOW FOIL

This is the second plane I invented for this collection that has reliably curved wing surfaces. This plane mimics the original hang glider design, which was actually first made as a spacecraft recovery system. The Rogallo Wing got the whole sport of hang gliding started, and you can still see remnants of the idea in the newest models. This harkens back to the first ones I saw growing up. The deep delta shape and billowing wings will give you great follow foil flying when you make this out of phonebook paper.

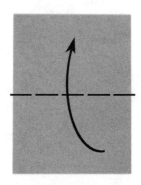

1. Start with the short side up. Fold the page in half.

2. The marked corner swings down to the center crease. The fold cuts through the upper right corner.

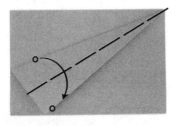

3. Fold down so the creased edge lines up with the raw edge.

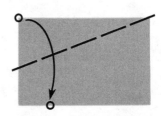

4. Flip the plane over and do steps 2 and 3 to the other side.

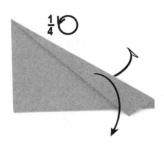

5A. Open up the center crease and rotate the plane a quarter turn.

5B. This photo shows the result of step 5.

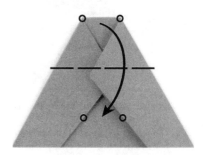

6A. This photo shows a closer view. Line up the raw edge with the two marked creased edges.

6B. This photo shows step 6 completed.

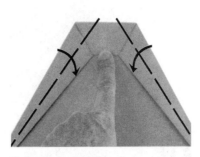

7A. This is zoomed out to show the layered parts being folded in half.

7B. This photo shows step 7 completed.

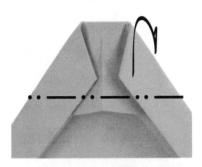

8A. This is zoomed in again to show the top getting folded behind. Leave about half of the raw edge layer on this side.

8B. This is still zoomed in to show the completed step 8.

8C. This is zoomed out. Look at the wings starting to curve! Flip the plane over.

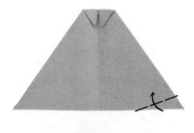

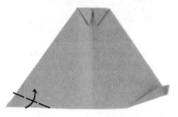

9. Make a giant up elevator fold. Use about half of the raw trailing edge.

10. Make the other elevator fold match.

11. Pinch the nose to get a little more curve in the wings. Try this as a glider. It's fun out of regular paper, but phonebook paper is the real trick. As I mentioned above, this shape is based on the old Rogallo hang glider shape. When you fly this as a follow foil, it's just like the old hang gliders using the updrafts along the coastal cliffs. It's precisely the same idea. The updraft counters the sink rate and keeps the aircraft flying.

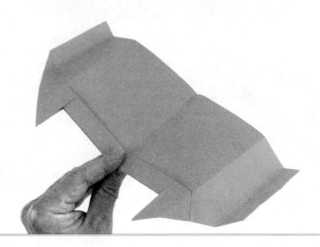

WALK THIS RAY

The triangular barbules (the small points jutting out in front of the wing) were the intriguing feature on this design. I'd heard that the air flowing around a point protruding from the leading edge could possibly swirl over the top of the wing to create lift. It seems to work on this design. You can adjust the angle of incidence up or down to improve the glide path. This is a very good follow foil when made from phonebook paper and a fun, if average, glider from standard paper.

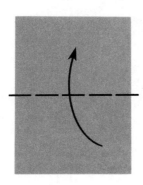

1. Start with the short side up. Fold the page in half.

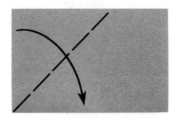

2. Fold the left edge against the center crease.

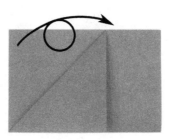

3. Flip the plane over and make the same fold on the other side.

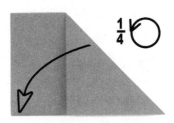

4. Open up the center crease and rotate the plane a quarter turn.

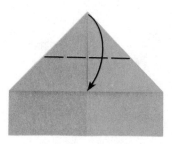

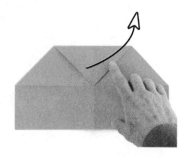

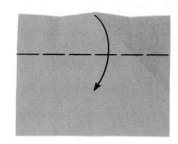

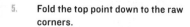

5. Fold the top point down to the raw corners.

6. Unfold all the steps.

7. Make a crease that uses the fold from step 5 and extends it to the raw edges.

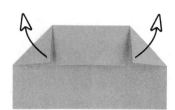

8. Fold the corners down, making diagonals across the small square shapes.

9. Unfold step 8.

10. Reverse the corners.

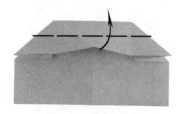

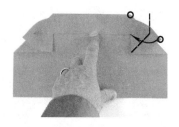

11. The top layer only gets folded in half. This will cause some weirdness.

12. Swing the right corner (the right-most circle) inside to where the layers meet (it should line up with the other marked circle). Pressing the side flat makes the marked valley fold happen.

13. Repeat the crazy step 12 on the left.

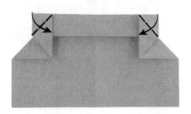

14. **Fold the corners down in preparation for reverse folds.**

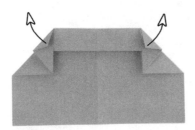

15. **Unfold the corners and prepare to reverse those points.**

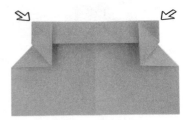

16. **Reverse these corners.**

17. **Mountain fold the back layer all the way across, but leave the barbules on the top layer where they are.**

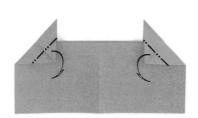

18. **Fold the corners under. The crease should be parallel to the creased edge of the barbules.**

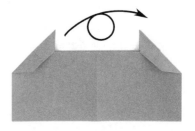

19. **Flip the plane over.**

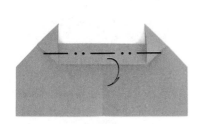

20. **Mountain fold the layers in half, tucking the lower edge underneath.**

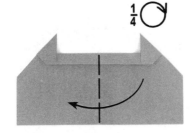

21. **Fold the plane in half and rotate the result a quarter turn.**

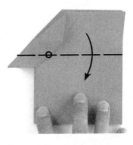

22. **The valley fold is parallel to the crease from step 21 and cuts through the layered corner marked.**

23. Fold up about one-third of the layer to make the winglet.

24. Flip the plane over and repeat steps 22 and 23 for the other side.

25. Unfold steps 21 through 24 so that the plane lies flat.

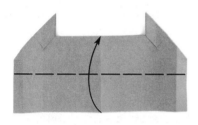
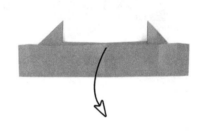
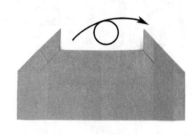

26. You should be looking at a mountain fold down the center. Fold the bottom raw edge up to the top layers.

27. Unfold step 26.

28. Flip the plane over.

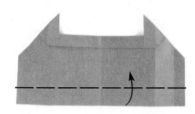
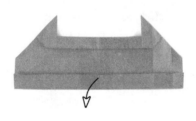

29. Fold the raw edge to the crease from step 26.

30. Unfold step 29.

31. Give the tail a pinch to create some up elevator.

32. The finished Walk This Ray. It's not much of a glider out of standard paper. It's a very good follow foil on thinner paper and can be flown quite easily around a room.

To fly this plane as a glider, you can reach under the plane and grip the leading edge. The back of your hand will face forward as you give the plane a gentle toss.

THE STORY OF THE WORLD-RECORD PLANE

5

Most people want to know how I got into paper airplanes. The truth is, I never got out of them. I'm an overnight success after more than forty years of folding and flying. If you want to know more about that, flip to the Backstory section, where I talk a bit more about my history with paper airplanes. This chapter is the story of the last three years, when I was trying in earnest to break the world record for distance.

Hubris. Hubris fueled by applause. This is how it all began. I'd done paper airplane shows for Google, Intuit, Genentech, schools, libraries, and science centers from here to Singapore.

I have a great collection of paper airplanes that do amazing things. *The Gliding Flight* and *Fantastic Flight*, which include lots of great planes, are books I'm really proud to have published. But here's the thing; I started believing my own hype. I started to believe that my paper airplanes might be the best in the world. What if I could get one of the best arms in the world to throw them?

In August of 2009, I finally lit the fuse: I started looking for a thrower to help me break the world record. After six months, I had already worked with two different quarterbacks—and the enormity of the task revealed itself. Using a stand-in thrower added complications. Their more powerful launches put stress on the paper airplanes that I hadn't imagined. Planes were nearly ripped in half from the violent acceleration. This meant that aircraft only lasted a few throws, leaving very little chance to get them adjusted before they had to be scrapped.

After six months, all I'd really managed to get was a firm grasp of the structural engineering challenge. Layering the paper to handle the throws was as important as the aerodynamics. After half a year, I had no winning plane, no permanent thrower, and no permanent place to practice. It was a slow start, and the finish line seemed to be on the other side of the galaxy.

LUCK AND TRAVELL

Finally, I lucked out finding a throwing space. Henry Tenenbaum, an anchor, reporter, and show host at KRON-TV had done a story on Airship Ventures, a company that does tours of the San Francisco Bay Area from a blimp (actually a dirigible). They fly out of Moffett Field (on the border of Sunnyvale and Mountain View, California).

I contacted Alex Travell, who was incredibly helpful. He invited me down to see the space. It's a blimp hangar, built to house airships bigger than the *Hindenburg*. In a word, PERFECT. Now I needed a thrower.

Vernon Glenn, a colleague, gave me Joe Ayoob's contact info, and that started the ball

Alex Travell at Moffett Field

The doorway of the leviathan Hangar 2 is more than 100 feet tall. The space is so large, it virtually has its own weather.

Me, admiring the height of the doorway, airplanes at the ready

rolling. There's an important lesson here. Don't keep your crazy dreams a secret. There's some risk to spreading the word you're trying to break a world record. Get over the idea you might embarrass yourself by failing, and just put it out there. Generally people like the idea of helping someone go big. Spread the word. You never know from what corner help will come.

THAT'S "BOOYA" SPELLED BACKWARDS

Joe Ayoob. In an era of hyperinflated sports figure egos, Joe is a refreshing guy to meet. The scouting report from Vern was Cal QB, arena QB, and paper airplane enthusiast. I was expecting some serious flaw when everything else seemed so perfect. Nope. He's in great shape; he doesn't carouse. He's in a steady relationship. He was giving tours at Anchor Steam Brewery when I met him and is now becoming one of their best sales reps, with a recent promotion to sales manager. At twenty-six, he had the quiet calm of someone decades older, and we shared the same childlike excitement over paper airplanes.

We had our first practice session at Moffett Field, Hangar 2, on July 14, 2010. Right away, things looked good. I'd been reworking a dart design. Basically it was a variation of the Tony Fletch plane. Tony held the record from 1985 to 2003, prior to

Joe Ayoob on world record day, February 26, 2012

My ballistic dart design. Note the notch to give Joe a better index finger hold. It flew 195 feet with pure brute force supplied by Joe.

Basic ballistics. Shooting a cannon or winning a pumpkin-throwing contest. Forty-five degrees is the magic number for the launch angle.

Stephen Krieger (who held it from 2003 to 2012). By the look of it, Stephen had also done a similar plane to Tony's. Since Stephen hadn't published, there was no way to be sure. Both Tony and Stephen had used a ballistic strategy, which calls for a low-drag design with just enough wing to maintain directional stability.

The launch is at a 45-degree angle, which is the optimum angle for punkin' chunkin' and cannon balls. Basically, if you're throwing an object that really doesn't fly, 45 degrees is your mark. You get the perfect parabolic path for best distance, for the force applied. If you streamline the object, you get better distance through less drag. This is a very straightforward solution to the distance problem.

On Joe's first day of throwing, we got to 140 feet. I added a throwing notch to the design. On his next day, we got past 160. After that, I started using vice grips to assist in making the creases. We got to 175. After that, I started using heavy iron plates to flatten the whole wing. We got to 195 (just past Tony's old mark of 193). We were scratching our way forward and thrashing Joe's arm with every practice. There were one or two small things I could still try. We were still working with an American equivalent paper stock. I didn't have access to 100 gsm A4 paper. I was working with 26lb 8½ by 11. A4 would add roughly 3.5 percent more weight, a distinct advantage for making a distance attempt.

Even though the plane was 3.5 percent lighter than allowed, and even though we hadn't tried absolutely everything yet, I couldn't see us beating Stephen's mark by much. The rules had changed since Stephen's throw, disallowing the 30-foot run-up. Now there's only a 10-foot space allowed, just a couple of steps. It seemed doubtful we'd break 207 feet by much, if at all. A prudent person might have stopped right there.

By now, the applause that had fueled my first days of effort had faded from my ears. My best dart ever isn't making the grade. I'm shredding Joe's arm with every practice session. I've got the best practice facility and one of the best available arms, and I still can't get it done. Perhaps this is the place to quietly bow out. But I couldn't. We'd come so far and had only 12 feet to go. I was hooked on the idea and couldn't stop.

THE IDEA

It was now January of 2011. I'd been watching and rewatching a video of Ken Blackburn and Takuo Toda throw their planes. The planes circle slowly for nearly 28 seconds. It's amazing video. You should Google it. It dawned on me that over the course of 28 seconds, those planes must certainly cover more than 207 feet. They had to. It's simple math. What if I just made a compromise glider, giving up a little wing area to stiffen the plane and making the fuselage a bit deeper to keep it tracking straight? Joe's arm can give us more than 50 feet of launch height. We'd just need a 4:1 or 5:1 glide ratio once it started to fly. I'd have to stiffen the wings, keep the center of gravity forward and low, and trail the layers as far back as possible to stiffen the fuselage. But yes, all that might be possible.

Guinness rules allow a maximum paper weight of 100 gsm (grams per square meter). The American equivalent would be roughly 26.4lb paper. Of course there's no such paper made. For a distance throw, you want all the weight you can get. Pushing against air molecules isn't as easy as it sounds. Guinness also allows tape: 30 mm of 25 mm wide tape.

So, where does one get 25 mm-wide cellophane tape? The same place you get 100 gsm paper. Not here. Mind you, I've been railing against adding tape to paper airplanes since 1989, when my first book came out. However, since leaving off the tape would mean leaving off weight, I decided to use the tape. Guinness allows you to cut that tiny tab of

A4 compared to US letter size

tape into as many pieces as you like. Eventually I chose to turn it into fourteen pieces. A quick bit of math tells you those are small slivers of tape.

THE RIGHT STUFF?

In late December, just in time for Christmas, Alex Travell had his wife send over some 100 gsm A4 paper. Mind you, Alex had been working with us for months now; putting us on security lists, escorting us back to the hangar, and watching us fly paper airplanes. His help and enthusiasm was indispensable. His wife, meanwhile, was in the UK. Perhaps she was happy to oblige his weekend forays into folded flight lest he occupy himself with other mischief. Whatever the case, she delivered the goods—about 120 sheets of 100 gsm laid vellum. Our world would never be the same.

So it was, that 15 months into this "slam dunk" project, I was starting over. I'd never invented a plane from A4 paper before I took on this challenge. How big a deal is that? It's still paper right? Well, yes, it's still flat and made from fibers. The bigger deal for folding is the paper's height to width ratio; A4 is taller than what I'd been folding with.

All my paper airplanes are made without measuring. One fold is referenced to another. Any small amount one place turns into a large amount somewhere else. Starting with a taller page meant that the center of gravity would likely be too far aft with any of my existing gliders. Most of my designs would simply be tail heavy. This turned out to be an advantage. I could add a layer to the front of the wing for stiffness and balance that by extending a layer partially down the body of the plane. There's a plan. Now just fold it.

I started with my basic design called the Phoenix. It's a bit of a crossover glider–dart design with a tall tail. I simply turned the diagonal folds into the leading edges of the wings. It took no more creases, and was aerodynamically cleaner. Good so far. I stayed with the old wing-folding technique I'd come up with for the Phoenix, moving the raw edge of the wing down to the corner of the fuselage. I like that move because it gives you a good-size wing and a deep vertical stabilizer.

It turned out that the taller paper gave me a longer layer to fold over before making the leading edge folds. This longer layer would help brace the fuselage. My very first try with the new paper looked pretty good. I gave it a toss. Wow! That thing tracks like a dream! It just locks onto a glide path and doesn't budge. You tell yourself that nobody gets that lucky. But there it is. The folding of this plane was to be the folding of the world-record plane. I would spend a year tinkering and, in the end, return to this basic idea with just eight folds.

THE WONDER OF RIDGES

Before I started to fold, I was confronted with paper that had ridges. Laid paper is placed or "laid" on the conveyor belt while still wet. The ridges get formed from the surface of the belt. I couldn't just throw it out, which was my first inclination. I imagined the ridges bulking up the thickness of my dart design, and who knows what it would do to the glider? And what about all that extra drag? This seemed like absolutely the wrong stuff.

A small ember started to burn in my brain after I watched my first glider made from the stuff fly.

Maybe the ridges help. Maybe they're like the veins in an insect wing; they create tiny eddies of turbulence that help the airflow smoothly follow the shape of the wing.

Suddenly I was in love with the ridges. Mrs. Travell, wherever you are, the Paper Airplane Guy thanks you from across the pond. My crazy dream now had ridges.

THE TALE OF ONE TAIL

On January 28, 2011, Joe had a chance to throw darts folded from the new paper. They were lackluster. They were thicker, and the extra drag ate up the extra weight advantage and then some. The dart idea was definitely stalled.

Then we tried the glider. Joe's first throw was 130 feet. I made a couple of tweaks and handed it back. Joe punched it out to 145 feet. We noticed the

A4 laid paper: A new beginning

This plane marked the first time I knew it was possible for Joe and me to break the record.

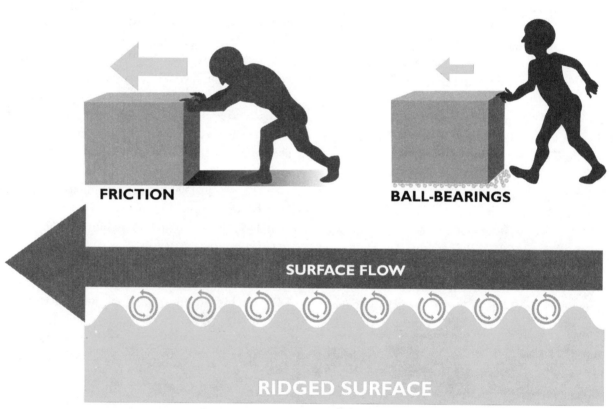

FRICTION

BALL-BEARINGS

SURFACE FLOW

RIDGED SURFACE

The eddies of air that set up in valleys actually act as bearings and reduce friction for surface flow.

back of the plane rattling, almost buzzing as Joe picked up the power. It sounded almost like a plastic bag held out of a car window when Joe threw it 165 feet. Amazing! We were onto something big. There were tons of things to adjust and make better. We could work on quieting the tail feathers. We could try different dihedral angles for the wings. We were free to experiment with launch trajectory and trim adjustments (tiny tweaks to the control surfaces). We were back in the flight game, not just "chunkin' punkins."

The very next day, Jay Hamilton-Roth, a local television producer, had scheduled an interview with me through a mutual friend. Unfortunately Joe had to bow out for this Sunday practice.

It was raining as Jay, Whitney, and I trudged across 150 yards of airfield to get to the hangar. The old wooden structure had a few leaks. Water pattered the floor here and there. I made a couple of tosses with the old dart designs and a couple of gliders I'd brought—just getting the fifty-year-old arm loosened up. I had, in my plastic carrying tub, a plane that I hoped would amaze Jay. I had changed my taping scheme to allow me to lock the tail together.

My first throw with the new glider went a full 120 feet. That was as far as I'd thrown anything, ever. I could see that this plane wasn't even adjusted well. I picked it up and carefully bent in a little up elevator. I gave it a hard toss—145 feet! Holy guacamole! I'm 20 feet further than my best ever, and plane is veering left and climbing too much. I smoothed out my up elevator tweak and nudged in a little right rudder. I took two steps and gave it a very hard fling—157 feet! That shouldn't be possible! I had almost equaled Joe's throw from the day before.

What could Joe do with this plane? My mind boggled. I gave the airplane a couple more throws before it finally wound up parked in a puddle of water, ruined. My best plane ever was trashed, my arm was thrashed, and I'd never been more certain the record would be ours.

It was straight math. Joe threw everything 65 feet farther than I did. If he did that with this plane, we'd be over the line! My best dart distance was shattered by 30 feet. If his best dart throw increased that much, we'd be over the line! We couldn't lose. It was a nice thought, anyway.

CLEAR AIR TURBULENCE

The Airship Ventures team hit the road, along with Alex Travell. It would be three months before we could get back into the space at Moffett Field. By then, Alex had taken a job back in the UK. With Alex gone, access to the hangar dried up.

With my paper supplies dwindling, my throwing space evaporating, a prize plane burning a hole in my mind, I did a desperate thing. I emailed NASA directly asking for official NASA permission to use the hangar. This was a bit dicey. Airship Ventures rented the space from NASA, and though nothing was ever said, Alex was definitely the kind of guy who would operate with a wink and a nod. Once you're on the official NASA radar, it's, well, official. You're definitely in, or you're definitely out.

Finally, in April 2011, I received an email from Stephen J. Patterson (ARC-JO) from the NASA–Ames Research Center, Aviation Management Office, which read in part, "We cannot allow you to use Hangar 2 for your project due to a number of legal and political reasons."

This was just the kind of baffling communiqué I feared. One might conjure up a legal rationale for denial of use—a slip and fall liability or such—but "political reasons"? Really?

Henry Tenenbaum, who used to be the assistant news director for Channel 9 in Washington, DC, enlightened me: "John, you just don't get it. First of all, you should've gone through their publicity department." This makes sense to me now, but at the time, not so much. "Some poor NASA person has to go before a Senate subcommittee and beg for funding. Some Senate staffer, trying to make a name for himself, will go over the Moffett budget with a fine-tooth comb," Henry said, closing out his email and turning to me to make his point. I still didn't get it.

"John, somebody is going to have to sit there and be asked, 'We see you have a line here for $5,000 . . . for supervisory and security time for . . . can you read this to me sir, because it looks like it says paper airplanes. I don't believe my constituents are prepared to support paper airplanes with taxpayer money.'" Henry raised his hands to signify surrender.

Oh, man. I knew immediately Henry was right. Political reasons! Thanks, in advance, Senate subcommittee dude. Nicely played. Was there no end to the political conspiracy to keep the record from me?

OF MAKER FAIRES AND MOJAVE

Well, actually quite a few things had conspired in my favor so far. This was a setback to be sure, but somehow we'd find another space—and one not controlled by the US government, its contractors, affiliates, or lackeys. But dang, they sure do have a nice collection of hangars I helped pay for.

May rolled around. It was Maker Faire time. A Maker Faire is like Burning Man meets county fair. It's the PG version of Black Rock married to some of the craziest, most ingenious backyard inventors you've ever seen. Somebody bailed on them at the last minute that first year, and I was a fill-in.

What can I say? There was a guy knitting with a pair of drumsticks, and he was using the knitting motion to play an actual set of drums at the same time. Deep, deep computer coding conversations about motion controllers and banter about the finer points of crochet happened at arm's length from each other. Three dimensional printers and a life-sized giraffe that gave rides to kids, CNC routers, RC Battleship wars, power-tool drag races, robot wars, and much more. It's just wonderfully crazy. I've been a regular exhibitor every year since.

This particular Maker Faire, Bob Withrow happened to see me performing, or more accurately, he was too far back behind the crowd that had gathered to see me perform. After the crowd cleared, he stepped forward and introduced himself.

"Hi, I'm Bob Withrow," he said, "I'm with Scaled Composites." Whenever a sentence ends with those four words, you should pay attention. Scaled Composites designs some of the coolest planes on the planet. I mention the Voyager, the Burt Rutan design that was the first aircraft to fly around the globe without refueling, in every presentation. Burt Rutan has been doing off-the-chart planes forever at Scaled Composites. Any

Me as reporter and presenter at Maker Faire, Bay Area 2010

Me and Bob Withrow, standing in front of Scaled Composite's Firebird—a reconnaissance plane that can be flown with or without a pilot on board

Here's a Burt Rutan design: Scaled Composite's Boomerang. Check out the shadow—asymmetry all the way.

time somebody dreams of growing up to be the most avant-garde aircraft designer the world has ever seen, they picture Burt Rutan in their dream.

"We'd love to have you come down and show us your paper airplanes," Bob said, extending a hand holding his business card.

"*The* Scaled Composites?" I asked, trying not to swoon. He just smiled.

"We just had a little friendly paper airplane competition with The Spaceship Company, and we lost," he admitted. "We don't like to lose."

Okay. Here's the deal. Scaled Composites is designing, building, and testing the prototype spaceship that will eventually take tourists into space. The Spaceship Company is building those vehicles. Virgin Galactic (owned by Richard Branson) will operate the spaceport (like an airport, only for space). And this guy just invited me down to Mojave, ground zero for pretty much the best aviation stuff on the planet. "Yes," I said, "All I want in return is a tour." He smiled and nodded.

THE DESERT DAYS BEGIN

This is how we came to throw paper airplanes in a spaceship hangar in the town of Mojave, located in the desert about 90 miles northeast of Los Angeles. After exchanging a couple of emails, I mention that I'm looking for a place to break the world record. Turns out, TSC (The Spaceship Company) is just finishing up FAITH (Final Assembly, Integration, and Testing Hangar) for the spaceship. We agreed I should look at the space during my visit.

The irony of the organization that once had the moxy to send a guy to the moon, now finding it too dangerous to let the Paper Airplane Guy make a few throws in their hangar, was just delicious. But then the private sector company, poised to take tourists into space, invites me to throw in their hangar? I mean, what are the odds . . . really?

It was the end of June 2011 when I made my first trip to wonderland. The town of Mojave is, well, Bob said it best: "The window of tolerance for most spouses is about seventy-two hours. That's when they realize where they've moved to and the tears start." Apparently being married to a cutting-edge aerospace engineer has its downsides.

What the town lacks in amenities, Scaled Composites and The Spaceship Company make up for in breakthrough ideas. And if you don't work there, I'm not sure why you'd hang around Mojave.

The presentation went well. The Scaled Composites engineers were amused by the variety of airplanes, and I taught them a killer distance design called the Stinger, from *The Gliding Flight*.

Bob gave me a personal tour of Scaled Composites that included their incredible spaceplane model, SpaceShipTwo, and the cargo plane used to launch it, WhiteKnightTwo. It was astounding.

Finally, it was time to test fly my own addition to aviation history. We followed the right-angled path of freshly made roads to the hangar. The concrete floor was still dusty from being surfaced. Other than that, there was no question you were in a brand-spanking-new aircraft hangar. The lights

Suzanne and me in the Scaled Composites lobby

First throws inside FAITH

were about 38 feet up and they had a zone motion–control system. The sound of relays clicking and lights flickering on followed you around the enormous floor. I set up to throw straight across the hangar, which would've been just a little short of the distance needed for a real record throw. We'd need to use the diagonal for the real deal.

I removed the top of the plastic container. You could see the paper start to curl, like a time lapse of a green leaf turning brown. Note to self: Acclimate paper in the desert before folding. I had a rough time getting anything to work.

I managed a couple of 125 foot throws, but the main thing was checking out the space. The ceiling was only 38 feet in a couple of critical areas. If you threw just right, you could squeak out 40 or even 42 feet dead center. It would be a little tricky but looked possible. And the cool factor was off the meter. Come on. We had to take a shot.

221 ON THE FIRST DATE

It was mid-July before Joe could break free for his first visit. We warmed up and made some throws. Whitney had joined us to record some throws, stretch the tape, and wrangle some planes. I was always glad to have him around. Everything is just easier with Whitney there.

The batch of planes for this trip had been folded in my living room with the fireplace roaring and the sunroom door open to let in as much warm air as possible. It was no desert, but it was as close as I could muster. It still took a full half-hour for the planes to start to behave.

Joe made some good throws right away—160, 170, 190, but he was still releasing a little high for my taste and causing a minor stall. I grabbed a plane and made a couple of test throws. Then I unleashed a monster while Joe was watching—200 feet! It had just the desired effect on Joe. He took the plane and threw it flatter than the surface of a lake. It was at 0.0 degrees. The plane lifted

The tape doesn't lie.

nicely, banked right slightly, slowed a little over the apex of the trajectory and then curved back left before settling into a straight tracking glide. It passed 100 feet still going up. It glided past 120 before it really hit its stride, 150 flew by as well as 200. It finally hit the opposite wall of the hangar about one foot above the floor.

The tape measure stretched to 221 feet and 6 inches! This was our first real proof of concept. We had just blasted past the old record by more than 14 feet! Bob Withrow was there and nodded in amazement. Joe, Whitney, and I laughed and cheered. It was as big a day as we could've hoped for. Joe's first throws with the new plane were beyond encouraging; they were world class.

It was a fun drive home—four and half hours of reliving that moment and dreaming of the world-record day inside a spaceship hangar. We were giddy. We couldn't know what was just ahead. Nobody could've predicted it.

PAPERS, PLEASE . . .

Problem #1: I had used up all of the original batch of A4 paper. When I wrote to Conqueror Paper to order more, they sent something with hardly any ridges at all. After much emailing, I finally sent a close-up shot of the ridges complete with the Conqueror watermark.

Uh-oh. I had an old batch of paper, so old it was impossible to match precisely. They checked warehouses and far-flung corners of offices to no avail. They finally found a blue paper stock that was a close, but not perfect, match. Would it work? It was our only hope.

Problem #2: Acclimating the paper. I sent some new paper stock to Mojave, in advance of the next trials. I would need to go down a day earlier to fold, of course. When I arrived, the paper had been fanned out, not placed as separate sheets, as I requested. Fanning allows the paper to bend slightly where it's not properly supported. It's a good idea for drying paper, if you don't plan on turning it into a flying machine.

Problem #3: We couldn't duplicate our record-crushing throw. With the new paper, our rattling tail feathers were back. Not as bad, but there. It seemed like a simple structural fix might do the trick. I changed the taping scheme to allow two pieces to hold the tail together. It helped, but there went a whole weekend on that idea.

FEELING THE HEAT

Next I tried a crease parallel with the center crease of the plane, all the way down the fuselage on either side. The plane still rattled on Joe's hardest throws, and now the chances of asymmetrical drag increased. If the new structural folds didn't line up perfectly, the plane would veer off course. You could correct this, but you added more drag by doing so. Once you start chasing your tail like this, it's frustrating. July was evaporating without great progress. August was bearing down on us, and beyond August, the spaceship was moving in. With the spaceship in, we'd be out.

Note the crease down the fuselage. This was one of several structural crease ideas tried.

LAYER IT ON

In early August, I came up with a novel adjustment to the underlayers of the wings. By adding just a tiny gap in the layering on one side, I could create just enough drag to slow that wing slightly.

If the plane turned left, I slowed down the right wing to get it to fly straight. This worked well for a whole session. We had several throws past 200 feet but none over the line. I still had a couple of tricks I wanted to try.

Opening up the right-wing layers to create more drag on that side of the plane

The result of adjusting the right wing

ILLEGAL U-TURNS

Armed with this new plane, we had a practice session two weeks before our scheduled Mojave record attempt in late August. The usual suspects were all there: Whitney, Bob Withrow, and Joe. With world-record day getting big in the windshield, we all felt a little pressure to make this work.

Sometimes, with everything on the line, you get a clutch performance that is simply sublime. This wasn't that. All morning and into the afternoon, plane after plane headed down the hangar, executed a U-Turn, and landed about 50 feet away. Some landed almost at our feet.

Bob Withrow was pacing the floor in sympathetic thought. At the moment, I was all out of theories. I asked Bob for an opinion. He murmered, "There's enough physics going on with that tiny wing for a doctoral thesis." He waited a beat before adding, "The air might be coming off the wing in different places at different speeds."

Okay. That's a big thought, and one heck of a problem. What do I do with that? How do I turn that into a feature? The answer wouldn't come to me for months yet.

There was one thing implicit in Bob's guess I *could* address. I could double my efforts at precise folding. Every part of the wing needed to mirror every part of the other wing . . . everywhere . . . to a level of high but unspecified perfection. That's a tall order. Hey, did I want it bad enough to try? I did.

One weekend of practice left. With a crazy folding regime for each plane, it now took a minimum of 30 minutes per copy. You shouldn't bother going without a dozen planes. Each plane might only last fifteen throws. Some less. Just when you get one working, it flexes wrong, breaking the spine, or crashes hard, wrecking a wing. Then you start over with the next plane. In short, you'd have to be nuts to try this.

We muscled through a very consistent practice: three planes past 200, with one at 205. With a blast of adrenaline and a little luck, Joe could crack this thing wide open. We still had a throw of 221 feet 6 inches in our memory banks. It was possible. We believed we'd do it.

A BIG MISS IN MOJAVE

August 29, 2011, arrived way too fast. There was a blur of tech set-up in the early morning. Three HD camera positions, the surveyor set up (since a tape measure over 200 feet long can stretch and is therefore not considered to be a "calibrated measuring device"), and a Phantom (super slow-motion) camera. This thing required so much light that we kept popping circuit breakers in the brand-new spaceship hangar. We were scrambling on the technical side to make good TV. On the paper airplane side, we'd had the planes overnight in the hangar. They seemed perfectly acclimated. I chose to hang them from their noses to keep the wings straight. That seemed to be working. Guests arrived. Judges arrived. Time clicked away.

CALM AT MACH 2.7

Mike Melvill was one of the judges for that day in August. He's the guy who flew the first privately funded flight to space, which resulted in him being awarded his astronaut wings. He also flew the first of the private-enterprise spaceship flights required to win the $10 million Ansari X-Prize (a competition to see who could be the first non-government organization to launch a manned craft into space). When I met him, all I could think about asking him was about that video of his flight. Google it and you'll see. Just as the ship really gets going, it starts barrel-rolling like crazy. There are shots inside the cabin. The lighting goes careening around as the spaceship rolls. It's a fixed camera, so there's no sense of motion except for the zany rotation of lighting around the cockpit.

Mike is a small, wiry guy, possessed of the calm that most test pilots have. There's nothing on the ground that can really rattle them, and few things in the sky.

"I have to ask you," I say, "what happened during that flight? Was it asymmetrical thrust? Did you lean on a control? What started that problem?"

Mike gave a little sigh. He answered thoughtfully, "I don't think it was a thrust problem. That motor has worked perfectly in every test and

The Mojave family contingent. From left: Susan Collins, Ted Collins, Adrianna Collins, Olivia Collins, Me, Ted Collins III, and Suzanne.

My nephew Ted and niece Olivia stand in front of Mike Melvill's ride to the hangar: a Thorpe T-18.

Me and Mike Melvill with the Long-EZ he built and flew around the world

The only place in the world they build this spaceship-carrying bird

seemed fine that day. I'm also sure I didn't bump the controls. You can see that in the video. Believe me, they looked." He smiled the smile of someone who's been to the proctologist and got a clean bill of health. "Our chief aerodynamicist thought the tail feathers might be a bit small. He told me to watch out for a little instability around mach 2.5. Sure enough, at mach 2.7 it performed an uncommanded snap roll and continued rolling for a total of twenty-nine vertical rolls before I managed to get it back under control."

I prompted him, "You looked pretty calm for a guy approaching mach 3 spinning that fast."

Mike gave me one of those test pilot mini-smiles, "If you look closely at the video you can see my hand start to creep toward the shut off. Then it occurred to me that if I shut it down, the spinning wouldn't stop—AND we wouldn't get the prize. So, I let her go."

That's pretty much all you need to know about Mr. Melvill. Nerves of steel at mach 2.7

and beyond. Making the right call while spinning inside a spaceship at close to 3 times the speed of sound. Yeah, that sounds simple, at least the way he explained it.

COMING UP DRY IN THE DESERT

August 29, 2011—a day that will live in infamy. In my grandest dreams of how this day will go, I envision me, making a miracle throw and breaking the record by a foot or two. For a moment, until Joe throws, I'm the best in the world. Then Joe crushes the record with authority. This is the dream I've played in my mind for a couple of weeks now—a moment of personal glory, followed by the definitive paper airplane distance achievement.

My first official throw is awful. I shank the thing into the hangar floor and it slides about 50 feet. Ugh. Throws 2, 3, and 4 net about 150 or

Hangin' in Mojave

so. They veer off course or stall. Throws 5 and 6 straighten out nicely and park around 165. Throw 7 is good. It sails almost 180 feet. Throws 8 and 9 are short of that. I line up for throw 10 and throw as hard as I possibly can. This is the full-out, leave-nothing-in-the-tank effort. The plane twists violently to the right and skitters an amazing 70 feet across the smooth hanger floor. There is a stunned silence and then polite applause. Phase 1 of my dream has just crashed on a bone dry floor in the Mojave desert. No problem. I've still got Joe.

Joe's first couple of throws are easily as good as my best. Throw 3 veers far left, impacting the hangar wall very high up, but gliding nicely as it does so. If I can straighten that out, we'll have a winner. Throw 4 goes better, but still too far left. Throw 5 is a beauty. It glides out 200 feet and 4 inches. I notice that it has stalled a bit at the apex of the flight. I caution Joe to release at just a bit lower angle. With a touch more power and bit lower angle in mind, Joe lines up for throw 6.

Throw 6 is a great throw. The power is good. The release is flat. It starts to climb, and climb and it looks beautiful . . . right up until it parks on an I-beam in the ceiling. This is a heartbreaker. We've got the plane really dialed in. We've got the throw working. And now, we're starting over.

Throw 7 hits a dangling extension cord that's been there for every practice session and never been in the way. The plane seems to have survived. I like the way it's flying, so we stay with it for throw 8.

Throw 8 spirals wildly to the right. I guess the plane was damaged by the impact.

Throw 9 goes 190 feet. This looks pretty good. We can push this one harder and get out there.

It's throw 10. Everything is on the line. With the trajectory this plane is flying, it feels like a safe gamble to throw level and go all out. Joe and I talk it over. All-in we decide. Keep the nose down a little so we don't bonk the ceiling, and just give it the juice.

Joe gives this plane a wing-rattling, fuselage-bending, monster of a throw. The plane torques hard left. It never has a chance. It slides a full hundred feet. His throw 10 exceeded my throw 10, but this was not the way I dreamed it: a hot, sweaty, fruitless end to our quest in Mojave. I slapped Joe on the back and shook his hand. I was crumbling inside as I approached the microphone and thanked everyone for coming. I felt as though I might faint.

It's the most awkward moment in my life. About 150 people lived the up and down of each throw that day with me. To be sure, it was gut wrenching for friends and family. I can feel the letdown in the polite applause, but it's not because they didn't see the record broken. They feel sorry for me. They're sorry I didn't achieve my dream. The pity was palpable.

Obviously we didn't break the record that day. The highlights are on my YouTube channel—just search for "Paper Airplane Guy." I made a throw of 178 feet. Joe topped out at 200 feet and 4 inches. That put him squarely between Tony Fletch and Stephen Kreiger. It's a world-class throw, but not a world-record throw.

We had a couple throws that are still the cause of debate in the Collins household. I contend that the throw that parked permanently on the I-beam could've beaten the record. I also believe the throw that hit the only dangling extension cord within city limits could've done it. We'll never know. We

Throw 6 lands in the rafters

Joe throws with calm determination and great form. I throw with desperation and a puffy face.

knew the space was tight. We knew we needed our best day ever. We flat out didn't get there. Even if we had, the flight wouldn't have equaled what we would eventually do at McClellan, a private jet center and general aviation field, near Sacramento.

Looking back, it's easy to see why the all-in throws failed. I forgot one very basic fact with paper airplanes. The harder you throw a plane, the more up elevator you need. The center of lift shifts back on the wing. Without adding up elevator adjustments, changing the release angle—that is, angling the throw more up—is the next best thing. I didn't do that, and I didn't tell Joe to do that. I was too busy with press, video crews, and mechanical details of the event. I failed to concentrate sufficiently on the most important purpose of the day: getting the throws right. I also feared hitting the ceiling, but logic instead of fear should have prevailed.

When it's time to break a record, you want a comfort zone. You want the margin—basically you want to have just an average day and get it done. Over our heads, pressed by a deadline, we took our best shot. We thought we could do it, but we failed. What I dreaded most was facing the sponsors. How could I tell them I spent their money and failed? Thousands of dollars is just gone, with nothing to show but some fancy video that wouldn't be watched. How could I face them?

YOU ONLY FAIL IF YOU STOP TRYING

It was a long ride home. Four and a half hours of recrimination. I felt like I let my team, my family, and my friends down. It was a heavy burden. I was out of fresh ideas, no place left to throw, and tired . . . really tired.

Inevitably Monday came as Mondays always do. I had work to hide behind for most of the day, but I knew I'd be talking to each and every sponsor before noon. I had nothing prepared.

It was at this point, I learned one of the most valuable lessons of the whole adventure: failing was not the end of the world. Not even close. Every sponsor—Fry's.com, XOJET, Airport Home Appliance, and Clark Russell—all simply wanted to know the same thing. "You're going to keep going, aren't you?" Business owners look at these things

differently. There's no shame in trying and failing. Further, you've only failed if you stop trying. Everyone said essentially, "We're with you. When will you try again?" It touched my heart.

August had been an emotional roller coaster. I lost a younger brother from a sudden heart attack. I hadn't really come to grips with that. The ups and downs of practice and progress, and then failing to break the record—I was exhausted from all of it. I know Suzanne, my wife, was ready for a break. She had been picking up all the slack for a while now.

A day later, Randy Fry of Fry's Electronics asked where we were throwing next. I told him that I was looking for another hangar. He said, "I've got one of those." I said, "How big?" He emailed a PDF of the layout. As I looked it over that afternoon, the old adrenaline started to churn. It was taller than FAITH. It was long enough to throw straight down the middle and not have to play the diagonal. It was in Sacramento, less than half the distance to Mojave. A practice was easily a one-day affair instead of a weekend. This looks doable! And just like that, I'd talked myself back into the game with no rest, no relaxation, and no consultation with my wife.

THE FRY DAYS BEGIN

Randy was overjoyed at the chance of hosting the record. Randy Fry's hangar is at the decommissioned McClellan Air Force Base.

McClellan, as I mentioned above, is now a private jet center and general aviation field. No commercial flights come and go, so it's fairly quiet. From September through December 2011, Joe and I spent nearly every weekend test flying. Joe and I analyzed the super slow-motion video from Mojave. He made adjustments to his throw, and I started working on the biggest issue: wing flutter.

It was obvious from watching the Mojave video: the harder Joe threw the plane, the more flutter we created, particularly at the trailing edge of the plane. We were burning up all the extra speed right there at the starting line. This explained the 200-foot wall we couldn't get past. Without fixing that, we could never hope to duplicate the throw of 221 feet 6 inches.

That throw now haunted us. We're still not sure how it was done. The perfect speed? A slight roll induced by a slightly wider wing? Exactly the right combo of wing imperfections? Whatever that magic bullet was, we couldn't find it.

Each practice session would include multiple variations on a concept, matched against our Mojave design. For most of September, October, and November, we stayed well short of 200 feet. Some days we were at only 170 or back at 160. These were the plodding, got-to-put-in-the-research sessions. We needed time to experiment.

Each weekend always started with great hope. The record barrier might be crossed with this crazy new idea. It progressed with throws that revealed another set of problems. The weekend concluded with analysis of the failure and a plan to modify the plane.

I wondered how long Joe would keep going. I needn't have. He always showed up, ready to try the next concept. He listened intently while I explained the theory and was quietly working out ways to improve his throw: a thumb adjustment here, an elbow down a little there, and a smoother acceleration all the way through. He really likes paper airplanes. He always threw the crippled planes around after the work was done, you know, for fun.

RESISTANCE AND RIPPLES

At this point, Suzanne was begging me to stop. With the dismal numbers coming out of practices, she started to dread the idea of me dragging all my friends out to McClellan to witness me fail again. It's a painful thing to put a friend through. I tried to assure her that we wouldn't officially attempt again until we were ready.

We definitely weren't ready. Frankly I was afraid I'd lose some enthusiasm from the sponsors if I backed off just to rest. That was probably unfounded. Finally Suzanne simply said, "I'm afraid you're going to put everyone through that again. Don't do that." "Don't do what—choke?" I asked. She nodded. After a pause, she added, "It's just too hard for everyone to watch. Don't do it until you're really sure."

In December, it finally occurred to me that the paper might be the problem. My pet theory about the ridges was almost certainly true for me. Was it true for Joe? Airflow on wings of this size is not well understood. Could it be that the very thing adding 30 feet to my throws was subtracting that much from Joe? Perhaps the ridges were the source of the high-speed fluttering.

SMOOTH IDEAS

I wrote to Conqueror Paper, requesting a selection of their stiffest and smoothest 100 gsm A4 paper. It was time to blow up everything and try again.

Conqueror Paper, it must be said, seemed as eager to get this done as I. They consistently delivered in a timely fashion, paid for the shipping, and gave me the samples. On top of that, it's a truly great selection of papers. They sent along about eight different paper stocks. I chose two that I liked and folded a couple of planes from each, along with a set from laid paper. Nothing fancy. Just straight copies of our Mojave design.

I started to notice that the smooth paper had a nice warp to the last third of the wing sometimes. It also had a sort of natural change of dihedral on some models, right where most wing layers ended. This was mostly from just folding, but you could definitely magnify it. Interesting. Noted.

This got me thinking about really optimizing the dihedral again. I skimmed my research one more time. The really interesting notion about dihedral angle is that, *if* you could build a plane that could change dihedral in flight, you'd really have something.

A lower dihedral is better during the high-speed launch phase. Less dihedral equals less drag. During the slower glide phase, more dihedral keeps the plane tracking straighter and nicely balanced. We're only talking about a five degree difference, but without some mechanical device and an actuator of some kind, how could you do it? I mean, it's still got to be just paper. Those are the rules.

Suddenly I remembered what Bob Withrow had said. The air was leaving the wing in different places as the plane changed speeds. I looked at the

A still extracted from the Phantom camera footage at Mojave shows how much the wings were deforming on launch: The left wing's rear corner is nearly vertical from the force of Joe's throw. Ridged paper helped me throw farther and burned up all Joe's extra strength with ripples.

natural warp in one of my planes. I just needed to flatten the nose dihedral and let it rake upward to the wingtips. It's already washing out toward the tail. The air will stick to the nose at launch and stick further back when it slows down. The dihedral will, in effect, change during flight. "Calm yourself," I thought. "It can't be that simple."

This was our last practice session in December. The temperature was about 50 degrees. Joe threw our old reliable Mojave designs. It felt good to be out in the 195-foot zone again, brushing up against 200.

We were nearing the end of the session, and Joe was really warmed up nicely. I handed him one of the smooth paper designs. It immediately felt good in his hands. "It feels stiffer," Joe said, "I like it."

On the first throw, the plane went 203 feet. That had never happened before. The next throw went 205. I looked at the plane carefully. It wasn't obvious with the ridged paper, but with the smooth paper I could see edges of the tape pieces now. That meant I could shave more drag by smoothing those down. I took a couple of throws. My best was 140. My distance had disappeared along with the ridges, but Joe couldn't seem to throw one short of 200.

We lost the planes with towering throws that ended up in the big hangar door and in the rafters, but we could both tell: this was big, very big.

2012 STARTS WITH A BANG (AGAINST THE OPPOSITE WALL)

I'd spent several days going through the paper samples again, finding the one with the best stiffness in all directions. I was also refining the dihedral-angle technique, flattening the angle at the nose, sweeping it up toward the wingtips, and letting it wash out again at the tail. I folded the few sample sheets I had into aircraft, and made backup planes from the second-best stock.

The temperature was around 53 degrees. The air was fairly dry. Plane 1 flew very well. It was immediately near 200 feet. On the fourth or fifth throw, it landed very high on the hangar doors, parked semipermanently on a ledge. It would've been an incredible throw, had it been straight.

We pulled another plane out of the plastic tub, made a few practice throws, and got ready for a ten-throw sequence. We decided that every practice this year would be like a real attempt. We needed to chart ten official throws, just for practice. Plane 2, on throw 1, flew 232 feet and 6 inches, hitting the opposite wall about one foot up. You can hear Joe whoop on the video. Again, check my YouTube channel—"The Paper Aiplane Guy."

We made three throws, inside of a ten-throw sequence, that broke the record that day. The new year was off to a very good start.

We would have five practice sessions in all before February 26, 2012, World Record Day. During each one, we broke the record by more than 20 feet.

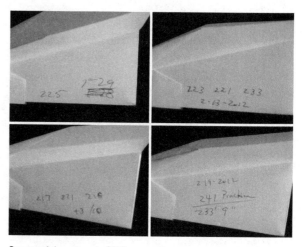

Successful pratices in 2012

Our best practice throw came just one week before world-record day. It flew an astonishing 240 feet.

We were heading into our second world-record attempt, and all we needed was just an average practice day.

DOUBT EVAPORATES

Coaching Joe's throwing had become like calling in precision artillery rounds. Joe could put the plane exactly where I'd aim him. "A hair to the left of the fire extinguisher, and just level with that white rail," I'd say. Joe heard and executed perfectly, consistently, and enthusiastically. This time there was no doubt. Suzanne's admonishment, "Don't choke," and the disaster in the desert were still fresh in my mind. But this time, we were ready.

This new idea for wing shape also spawned a new idea for aircraft adjusting. I was now making adjustments to the front of the plane if it veered off course on launch. In addition, if it curved after it slowed down, I could adjust the tail.

Joe winds up

Note the toe placement. Joe gets it right every time.

The distance we had to cover: just a bit over 75 yards, wall to wall

Generally, at midspeed, I let the plane fly freely. Sometimes I'd flex a wingtip to get a better balance. I now saw adjusting the plane as a dynamic process that needed to be matched to the problem *and* the speed of the plane.

I was all the way back to the original first model I'd folded with A4 paper: just eight well-made creases. True, there was a complex taping scheme, a dihedral-angle scheme, and many other things. I now "candled" the paper before folding, holding it up to an incandescent bulb and searching for interior flaws in the paper. Sometimes there's a structural blemish you can't discern with the naked eye. I also warmed the paper slightly to dry it and confirm which direction the paper naturally curled, so I could take best advantage of my wing warp.

I turned the downstairs storage room into my lab. We jokingly referred to it as "the global headquarters for the Paper Airplane Guy." Using space heaters, I kept the room at 78 degrees around the clock. The paper was always the same temp and humidity. Only I entered the room. Obsessive? To be sure. I wanted to crush this record now.

Everything that I could do would be done in that pursuit. We were so close now. We just needed one more good day—actually just an average practice day would do it. I just had to stay focused.

FEBRUARY 26, 2012: WORLD-RECORD DAY

Friends and family gathered. Whitney Alves was fully in charge of all the cameras and media wrangling this time around. I did no press interviews before the throws. I had only one thing in mind: make this like any other practice day. To that end, we were determined not to rush. Pick the right plane, make the right adjustments, wait to get past 200 feet, and then take a shot. That was the plan.

At 9:45 a.m., two L-39s in full Patriot Jet Team colors did a fly-by, smoke trails and all. It was a glorious start for the day.

Ken Blackburn, four-time world champion for duration, had agreed to be a judge for the event. The other judge would be Stephen Kreiger, the

The crowd assembled at McClellan Air Force Base

A box of practice planes, one of which executed a loop before flying past the world record mark during a practice throw. It was fluke, but incredibly fun for the audience to see.

Joe looks on as I make an adjustment.

Henry Tenenbaum, who emceed both record attempts and took photos

current world-record holder for distance. I wanted no doubt at the end of the day that we'd broken his record. If Stephen agreed we had, how could Guinness not certify us?

I had a bunch of practice planes left over from earlier in the week. I'd done live shots with local television stations all week leading up to the event, so I had plenty of planes that were too old to actually be thrown for the record.

One such plane was launched, did a loop more than 40 feet in the air and proceeded to glide all the way past the world-record mark, which was lit with rope lights on the hangar floor. When Sal saw the plane go over the line, he started flashing the lights. The gathered faithful roared their approval. Henry and Vicki reminded everyone that it was only a practice throw. This set the stage perfectly.

Now everyone knew what to expect when a plane breached the world-record barrier. Joe and I shared a laugh. That was a first. We'd never seen that before. No plane had ever looped and come close to the record. Neither one of us thought we'd ever see it again.

We would plunk one more throw past the record mark before I decided to have Joe make his first official throw. Throw 1 landed well short. Ken Blackburn decided to place himself at the finish line to ensure accuracy in marking the landing. Stephen stayed at the start line to ensure compliance within the throwing area.

Throw 2 was very close to the record line. Throw 3 veered hard left. A lot of practice throws went left that day before we could get a good shot down the center line. It turned out that someone had left the back door to the hangar open, and a

This still, taken by Anton Wannenburg, is the winning throw crossing the world-record barrier. Sure, it's fuzzy, but it's the one still shot of that moment. Hey, you try guessing where that plane will cross the line.

very slight diagonal breeze was drifting to the left, escaping through the door. The incomparable hair stylist, Clark Russell, claims to have noticed hair literally being blown by unseen winds and passed the word up the chain of command. I believe Jack Shiefly closed the door before the official throws began.

We took a couple more practice throws. One crossed the world-record line. It was time to nail this thing. Believe me, you don't want to be looking at throw 9 or 10 in a do-or-die situation. I did that in Mojave. Trust me. You die. I scooped up the plane, reasserted my nose tweak and a little up elevator that this plane needed. I walked it back to Joe and nodded. "Let's do throw 4."

We were angling the throw to the right to compensate for an early veer to the left during launch. The plane had a tendency to drift back right after it started to glide. So a little left rudder was called for.

If you'd asked me how to adjust this plane, say six months ago, for the same issues, I'd have given you the wrong answer. I'd probably have bent in some right rudder and pulled down the underlayer

The crowd approves!

on the left wing. Way wrong answer. That's a recipe for a U-turn!

Joe nailed the throw, releasing just barely above horizontal. I knew immediately we had a good throw. As the plane neared the halfway point, it was tracking straight, having recovered from the right-of-center launch. When it tipped over the top, I knew it was going to flare nicely.

Souvenirs from the day: Ken Blackburn, the four-time world champion for duration, folded and signed a plane for me. Not to be outdone, Stephen Kreiger folded and signed his world-record design. Two of the best in the world joined me for breaking the record. It doesn't get much better than that.

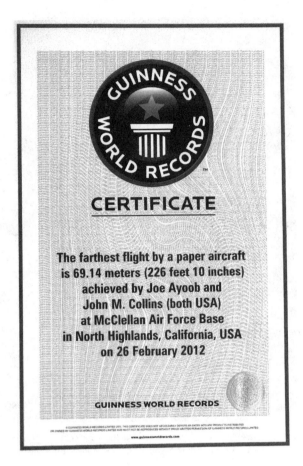

GUINNESS WORLD RECORDS

CERTIFICATE

The farthest flight by a paper aircraft
is 69.14 meters (226 feet 10 inches)
achieved by Joe Ayoob and
John M. Collins (both USA)
at McClellan Air Force Base
in North Highlands, California, USA
on 26 February 2012

GUINNESS WORLD RECORDS

From left: Viki Liviakis, Randy Fry, Clark Russell, Me, and Joe Ayoob in front of an L-39 jet

We had a record breaker, but by how much? The crowd could sense it too. If you listen to the video, you can hear the cheering build as the plane gets closer to the line. They explode as Sal flashes the rope lights. I wait to yell until the plane touches down. After verbally coaxing it down to the finish line, I hold my breath. It ain't over 'til it's over. If the plane touches anything or anyone before it hits the floor, that's where the flight is judged to have ended.

Finally, after flying gently over the record mark, and floating another 19½ feet, the plane gently slid to a stop on the hangar floor. The rope lights were flashing, the crowd was roaring, Joe picked me up and swung me around. The whole place went nuts. Confetti cannons were fired amid the backslapping and handshaking.

Ken Blackburn supervised the marker placement for the official surveyor shot. I found my wife and held her tight as I swung her around in a circle.

We took six more throws, of course. One landed on top of a light 42 feet up. That had never happened before. One banged into a light on the way up. That had only happened once before. The others were just a bit short.

It didn't matter. We had won. We finally won. My wedding day, the birth of my son, and world-record day—the three biggest days of my life. What a long, amazing journey from twenty-three years ago when my first book was published to this moment.

THE BACKSTORY

First, a little history.

I was born in 1960 in Eureka, California, on the date that in three years' time no one would forget—November 22, the day of the John F. Kennedy assassination. I own my birthday for exactly 2 years and 364 days. After that, it is a national day of mourning.

It is noted by my kindergarten teacher that I like recess and riding the tricycle. These appear to be the only noteworthy achievements for my very earliest "permanent record."

At some point in the third grade, I make my first paper airplane. It's the classic dart. I make a lot of them. Even here there is controversy. My older brother, Ted, folds the paper in half and leaves it folded. He folds the corners down, and continues folding two more creases in the same direction. My mother's method differs slightly. She unfolds the lengthwise crease and folds the corners to that crease. Then she remakes that crease with the corners now inside rather than outside the main crease.

I find Mom's method superior to my brother's, even though the top of the wing is not smooth on the finished plane. Fewer layers flop open in flight. Ted lobbies for his method and loses my vote. This annoys him and emboldens me.

In the fourth grade, my mother shows me a paper airplane design that involves an origami move known as a waterbomb base. We certainly didn't know it as that. This base informs my folding for decades to come. It's still a mystery how she knew how to fold this plane. Were there days of misspent youth she was not divulging?

Later this year, my older brother lifts the basic weight-creation system from Mom's plane and leaves the strip of paper attached—the paper that usually becomes the tail—and puts in a stair step-shaped wing fold. It's a superior design in every way. It becomes the first paper aircraft to cross

Mom's design. It's a sort of classic nose-weight system that starts with a waterbomb base.

Social studies worksheet after grading. Sweet revenge. This is a copy of my first design to fly completely out of sight, circa 1971.

Chaffin Avenue, float to the height of more than two telephone poles, and fly completely out of sight. I'm jealous beyond description. I make it a point to immediately learn to fold it better than him. I eventually duplicate the flight two weeks later.

The paper airplane arms race is on, at least in my mind. Meanwhile, his attention has suddenly turned to a full-time investigation of the girl who just moved in at the end of the street. I view this as a strategic blunder on his part.

One month later, I develop an odd variation of a sink fold that involves making two small cuts in the paper. It allows me to create symmetry on the layer above and below the wings. It barely limps across the living room indoors. I take it outdoors for a more thorough test.

On throw 4, it catches a perfect gust and rides very high across the road. It's already well above the power lines as it circles upward on the near side of the field. It climbs higher and eases its way toward the row of spruce trees on the far edge of civilization. It circles up and over and then out of sight. There's instant jubilation! My heart is pounding hard in my chest. A lump forms in my throat.

Then, suddenly, a startle reflex crackles through my whole body. That was my prototype! I race indoors desperately hoping to remember how I folded that plane. The need to document my work is born.

In fifth grade, a substitute teacher decides that bringing an origami book into the class would be an easy time-filler for art class. It's 1970 in Humboldt County, California. There is no origami paper within 250 miles. This teacher had obviously stayed up all night cutting colored letter-sized paper into squares. At this point in my folding career, checking to see how square a piece of paper is, is child's play. One diagonal fold does the trick. After I point out the umpteenth unsquare piece, she admits her folly.

After attempting to teach a bunch of fifth graders a "simple" model (the crane), she may have given up folding entirely. At recess, she leaves the origami book on one of the craft tables. I pick it up and enter a brand-new world. Moves I thought I had invented were hundreds, if not thousands of years old and done so much better. It is the first book I ever ask my parents to buy.

I begin to dream of writing a book of my own paper airplane designs. I already have a shoebox full of original designs. What could stop me now?

When I'm in eighth grade, the Kline–Fogleman wing gets major play on a CBS morning news segment geared toward kids. Christopher Glenn narrates the piece about an amazing stall-proof wing, originally designed as a paper airplane. The paper airplane design has a step, or discontinuity, on the bottom of the wing. This stepped shape under the wing creates unexpected advantages. Geeze . . . there's real competition out there. The 1st International Paper Airplane Contest book was out and I was working my way studiously through every design. But this? This ups the stakes. People apparently would take this kind of work seriously if you were really onto something. Was I? Could I? More folding and testing would certainly be needed.

High School. My folding goes underground. Paper airplanes do not make you cool. I'm not sure what, outside of breaking a bone during a football game, does. I'm not willing to do that.

Today I have a job, a wife, a son, and other hobbies. I do, however, have a major paper airplane habit. This obsession has made it possible to travel to Austria; Singapore; Bali; New York; Chicago; Portland, Oregon; St. Louis; New Jersey; Spokane; Seattle; and numerous San Francisco Bay Area cities—all at the host's expense—just to fly paper airplanes.

I'm fortunate to have performed for Intuit, Google, Genentech, and XOJET without leaving the Bay Area. Maker Faire Bay Area and Hiller Aviation and Space Museum consistently book shows, so if you're traveling to the San Francisco Bay Area, check their calendars.

My current job is being executive producer and creative director for PD TV, a division of Petersen Dean Roofing and Solar that creates commercials and a TV show called *Saving Green with Petersen Dean*. Before that, I was an on-camera host and voice-over talent for a television station, at which I also produced, directed, and wrote commercials and infomercials. Before moving into commercial work, I was a director of live newscasts for more than twenty years and for part of that time was the news production supervisor.

I love to windsurf. Most weekends during the windy season, I'll be ripping across the water at

Larkspur Landing, north of San Francisco. Sometimes I'm one of those crazy windsurfers you see near the Golden Gate Bridge. If you want to learn about aerodynamic forces in a very tactile way, I highly recommend the sport. The concept of a quadrupling of force with only a doubling of airspeed is readily understood as you go head over heels into the water from being blasted by a monster puff of wind. Bliss!

I have a relatively small collection of boomerangs. I'm fascinated by the physics there too. Lift, precession, centrifugal force, and more. What a sport. There are left-handed and right-handed 'rangs by the way. I didn't realize that until I ordered one wrong. Oops! Makes perfect sense after you think about it.

I love to swing dance with my wife, Suzanne. We're not very good, but we have a lot of fun trying.

About the only other hobby I've seriously pursued that lacked an aerodynamic component is origami. It has direct application to my paper airplane efforts, but it's really a pure art and aesthetic in its own right. I can't tell you exactly how this hobby will benefit you. All I can tell you is that the hours you spend folding paper will help you achieve something else in your life. You'll learn patience, if nothing else.

I loved math and science in school and studied liberal arts as a college student. One of my great goals is to foster a passion for the pursuit of knowledge. The world is an amazing place, made more so by discovering the workings of everyday things. Flight, gravity, and light are really still mysteries waiting to be unraveled. We understand a good deal about how to use these things, but the true nature of them remains elusive. My advice? If someone tells you all the really cool stuff has already been discovered, stop talking to them immediately and seek better company. A closed mind will never open the mysteries of nature.

"Settled science" is hardly ever that. Science is always teetering at the peak of what we can prove. That idea makes some people uncomfortable. To me, it's like windsurfing. I can't control the wind, the tide, or the waves; the best I can do is come into balance with all those disparate forces and hang on to that joy while it lasts. A good theory is like a great ride on the board. It's really satisfying, and there's probably a better one waiting just ahead.

SPECIAL THANKS

To my sponsors, Frys.com, XOJET, Petersen Dean Roofing and Solar, Clark Russell Salon, and Airport Home Appliance, thank you for your generous support. In particular

Randy Fry for the hangar, the Patriot Jet Team, the pizza on game day, and your tireless support of the whole effort.

Stephen Lambright for convincing XOJET this was a good idea.

Jim Petersen for stepping up and Keith Nash for showing up to snap a few photos.

Clark Russell for just being fabulous.

Paul Myer for saying yes to sponsoring before the question was completed. Alicia Owsley for making sure the shirts were ready, twice!

To the major players in the drama

Bob MacDonald for the Sultan of Oman story, and others.

Jack Schiefly for moving jets large and small, and making sure we could fly in the hangar.

Art Takeshita for always having the right tool and the right shot.

Dean Kendrick for wrangling cards, cameras, and the killer center camera shot in McClellan.

Rico Corona for all the right toys and expertise to use them.

John Chater for super slow-motion work beyond the call of duty.

John Fontana, the right guy with the right lights and a smoke machine.

Whitney Alves, whose boundless energy made the whole thing more fun. From shooting to editing to finding more rope

lights, I can't imagine having done it without him.

Sal Glynn, my first book editor and long-time friend.

Jan Adkins for all the great illustrations in this book. He's the explainer 'n' chief and principal wizard at the Jan Adkins Studio. He insists on calling me Captain Tomorrow, today; it's all very confusing.

Jim Swanson, my ex-boss, my friend, and fellow flight fanatic. Can I borrow a stick of Beeman's?

Alicia Cocchi, my coworker and event photographer. Thanks for snapping the right shots.

Anton Wannenburg for the shot of the plane crossing the line.

Henry Tenenbaum for his amazing photographs and even more amazing emcee work at both world-record attempts.

Vicki Liviakis for introducing me to Randy Fry and Clark Russell, and for emceeing with Henry at McClellan.

Alex Travell for making Moffett happen and giving us our first taste of A4 paper. Mr. and Mrs. Travell will always be large in the story of this paper airplane.

Wayne Whatley, our surveyor in Mojave. Many thanks for a job well done.

Dirk Slooten, our surveyor at McClellan for the record breaker. Great work!

Tim Piastrelli for website work above and beyond what mere mortals can accomplish. Thanks for all your help.

Bob Withrow for inviting us to Mojave, inspiring the solution, and putting us up for a night at your house. What a guy!

Enrico Palermo for offering up FAITH for the Mojave attempt. Thanks for all your help and good wishes.

Mike Melvill for being a judge at the first attempt. An amazing and inspirational guy.

Willie Turner and Hiller Aviation and Space Museum, San Carlos, California, for letting us shoot stories and host television live hits from the museum. Thanks.

KRON-TV for generously allowing the rental of their gear for the second and successful record attempt.

Tim Martin for being Mr. iPhone app.

Richard Zinn for putting me in touch with Mike.

Vern Glynn for putting me in touch with Joe Ayoob.

Mike Pawlawski for giving it a try.

Jimmy Collins for ripping my planes in half. No really . . . that was helpful.

Rip Wong for letting us use the Mercy Hangar with no advance notice.

Ken Blackburn for traveling all the way from Florida to be a witness for the McClellan attempt. Four-time world champion and endless amounts of class.

To my family

Ted and Olivia for chasing planes for miles at a time.

My brothers—Bob, Ted, and Jim— for providing competition, inspiration, consolation, and celebration. Thanks, guys. I could feel you there even when you couldn't make it in person.

My son, Sean, for showing up.

My sisters-in-law, Faith, Robin, Susan and Becky—like the sisters I never had.

My wife, Suzanne. Without her love and support, none of this would've been possible. Now she's the champ in name as well as in my heart. I love you.

And Joe Ayoob, the best paper airplane thrower in the world. I knew you could do it all along.

FOLDING SYMBOLS

—— —— —— —— —— —— VALLEY FOLD

•• —— •• —— •• —— •• —— MOUNTAIN FOLD

———————————————— EXISTING CREASE

FOLD THIS DIRECTION

UNFOLD THIS DIRECTION

FOLD AND UNFOLD

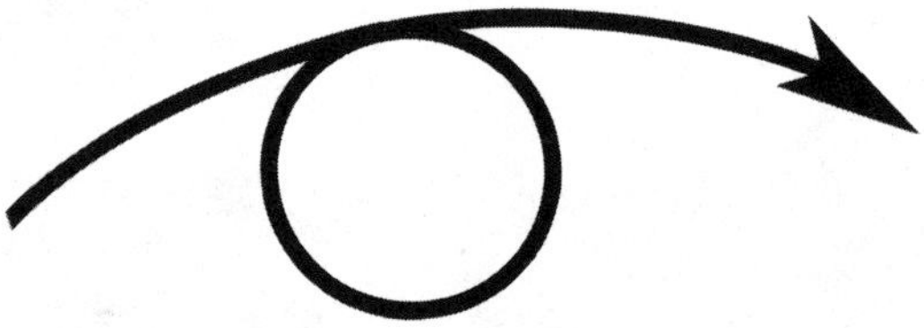

FLIP OVER

FOLD BEHIND

WATCH THIS POINT

SINK, SQUASH, REVERSE, OR PUSH

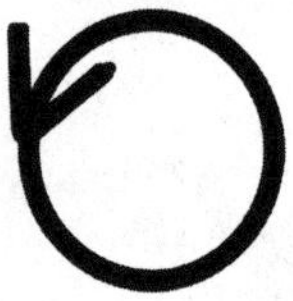

ROTATE THE PLANE

Suzanne, page 33

Max Lock, page 56

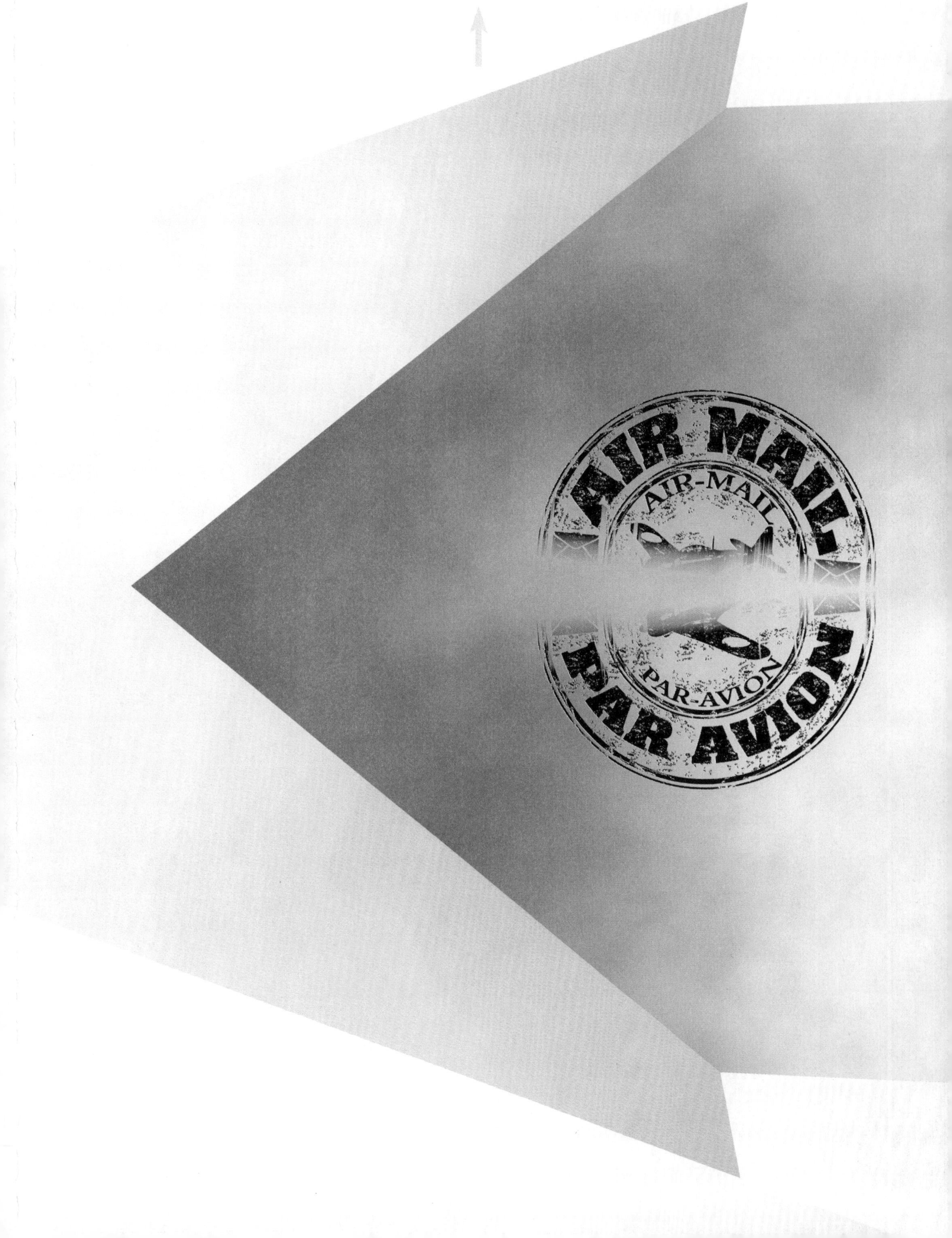

Max Lock, page 56

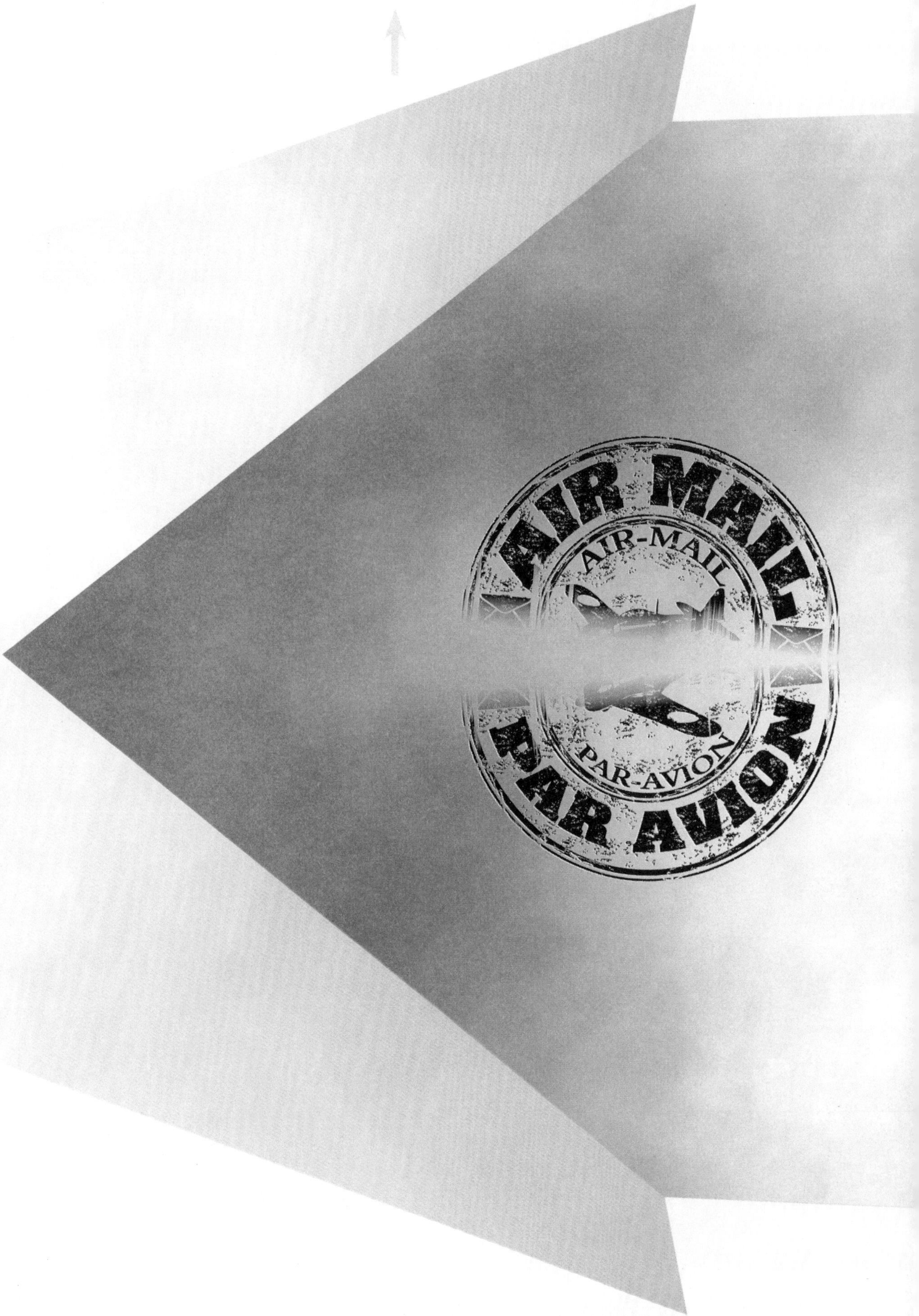

Locked and Loaded, page 62

Javelin, page 40

Javelin, page 40

Ultra Glide, page 64

Ultra Glide, page 64

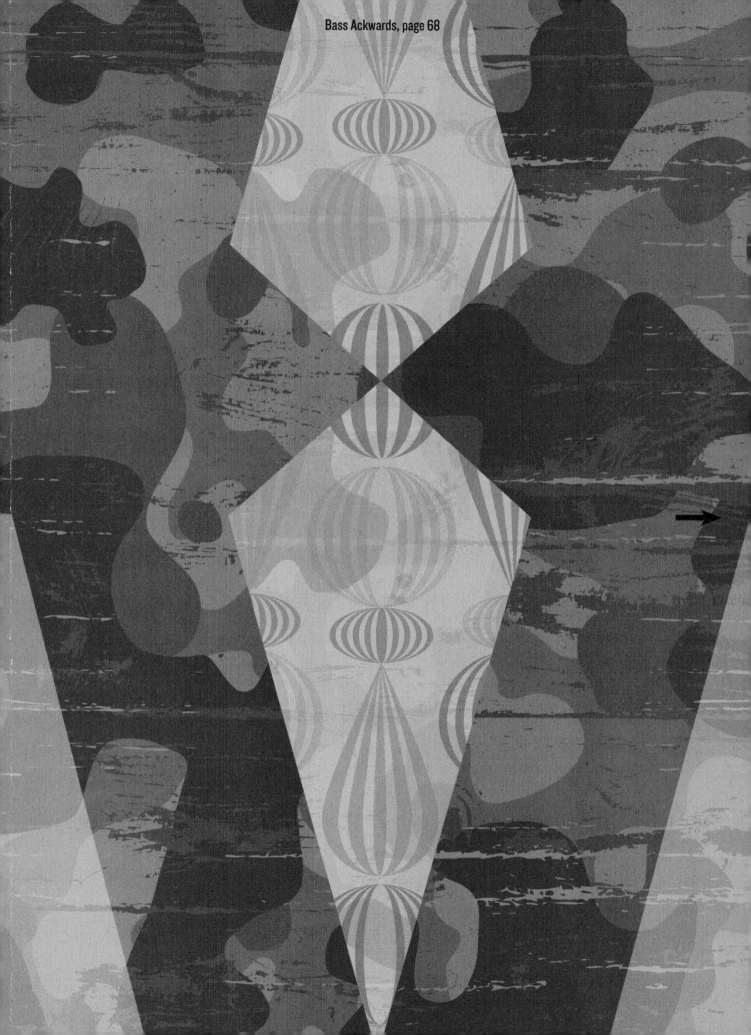

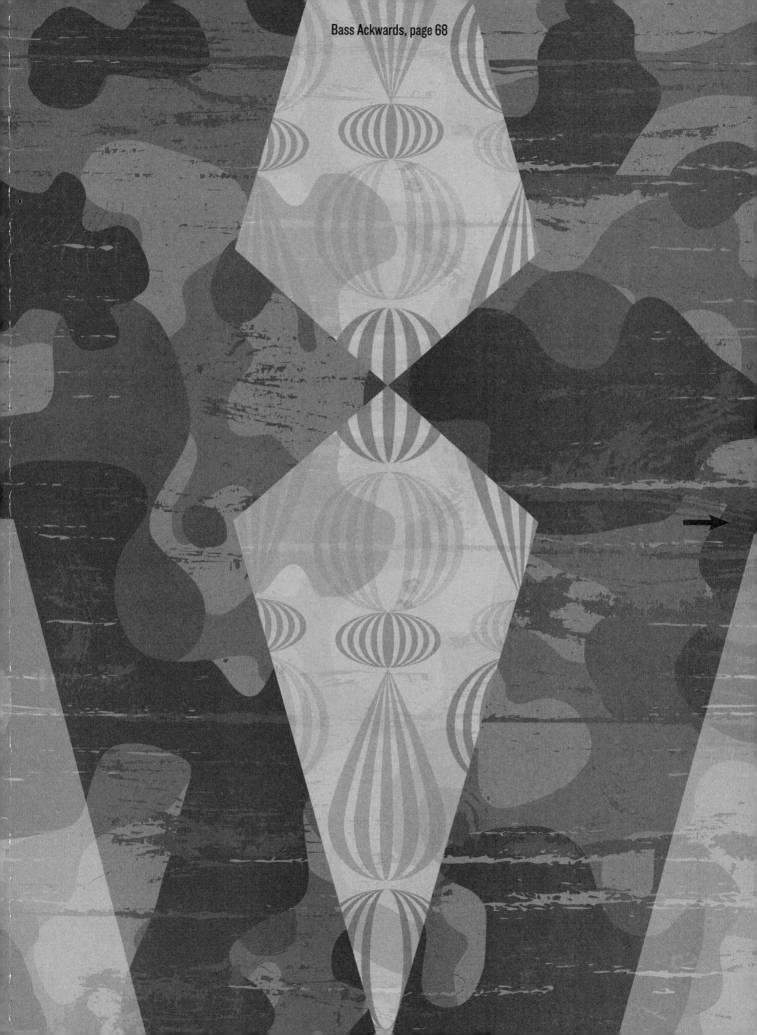

Pro Glider, page 46

Pro Glider, page 46